THE OPEN UNIVERSITY
Science: A Second Level Course

Block 3 Part 11
Ore deposits 11 : exploration and extraction

Prepared by an Open University Course Team

S238
THE EARTH'S
PHYSICAL RESOURCES

The Open University Press

The S238 Course Team

Chairman
Ian Gass

Course Coordinator
Isla McTaggart

Authors
Geoff Brown
Steve Drury
Graham Jenkins
Patricia McCurry *(Consultant)*
Dave Park
Julian Pearce
Eric Skipsey
Peter Smith
Sandra Smith
John Wright

Editors
Dick Sharp
Jacqueline Stewart
David Tillotson

The following people have assisted with particular parts or aspects of the Course

Andrew Crilly *(BBC)*
Sir Kingsley Dunham *(Consultant)*
Dee Edwards
Janis Gilbert *(Illustrator)*
Laurie Melton *(Librarian)*
Pam Owen *(Illustrator)*
Jane Sheppard *(Designer)*
John Simmons *(BBC)*
John Taylor *(Illustrator)*
Graham Turner *(BBC)*

The Open University Press, Walton Hall, Milton Keynes, MK7 6AA

First published 1984.

Designed by the Graphic Design Group of the Open University.

Filmset and printed by Thamesdown Litho Limited, Swindon, Wiltshire.

ISBN 0 335 16158 8

This text forms part of an Open University course. For general availability of supporting material referred to in this text please write to: Open University Educational Enterprises Limited, 12 Cofferidge Close, Stony Stratford, Milton Keynes, MK11 1BY, Great Britain.

Further information on Open University courses may be obtained from the Admissions Office, The Open University, P.O. Box 48, Walton Hall, Milton Keynes, MK7 6AB.

1.1

Contents

Study Guide and Table A for Block 3 Part II

EXPLORATION AND EVALUATION

1	**Exploration strategy**	**6**
1.1	Exploration, finance and risk	6
1.2	Summary of Section 1	8
	Objectives and SAQs for Section 1	9
2	**Geological exploration**	**10**
2.1	Defining the target	10
2.2	Prospecting	11
2.3	Remote sensing	12
2.3.1	Aerial photography	12
2.3.2	Multispectral remote sensing	13
2.4	Summary of Section 2	15
	Objectives and SAQs for Section 2	16
3	**Geochemical exploration**	**16**
3.1	Geochemical dispersion	18
3.2	Geochemical surveys of soils	18
3.3	Geochemical surveys of drainage	20
3.4	Vegetation surveys	20
3.5	Planning a geochemical survey	21
3.6	Summary of Section 3	23
	Objectives and SAQs for Section 3	24
4	**Geophysical exploration**	**24**
4.1	Gravity surveys	25
4.2	Magnetic surveys	28
4.3	Electrical surveys	30
4.3.1	Resistivity surveys	30
4.3.2	Spontaneous polarization surveys	32
4.3.3	Induced polarization surveys	33
4.3.4	Electromagnetic surveys	33
4.4	Summary of Section 4	34
	Objectives and SAQs for Section 4	34
5	**Evaluation**	**35**
5.1	Geological considerations	35
5.2	Logistic considerations	37
5.3	Summary of Section 5	38
	Objectives and SAQs for Section 5	38

MINING, MINERAL PROCESSING AND SMELTING

6	**Mining methods**	**39**
6.1	Choosing a mining method	39
6.1.1	Mining cut-off	40
6.1.2	Stripping ratio	42
6.2	Surface mine design	45
6.2.1	Open-pit or bench mining	45
6.2.2	Open-cast mining	47
6.2.3	Alluvial and marine mining	47
6.3	Underground mining	48
6.3.1	Access	48
6.3.2	Mines in shallow-dipping ore deposits	49
6.3.3	Mines in deposits with large vertical extents	50
6.4	Summary of Section 6	52
	Objectives and SAQs for Section 6	52
7	**Mineral processing**	**53**
7.1	Basic operations	54
7.1.1	Mineral identification and distribution	55
7.1.2	Siting of processing plant	55
7.2	Principles of mineral liberation	55
7.3	Comminution	56
7.4	Sizing the output from comminution	57
7.5	Mineral separation	58
7.5.1	Mechanical sorting	59
7.5.2	Float–sink methods	59
7.5.3	Shaking methods	60
7.5.4	Magnetic separation	61
7.5.5	Electrostatic separation	61
7.5.6	Flotation	61
7.5.7	Leaching	62
7.6	Choosing a separation system	62
7.7	Drying procedures	64
7.8	Summary of Section 7	64
	Objectives and SAQs for Section 7	65
8	**Metal smelting**	**66**
8.1	The stability of metal ores	66
8.1.1	Metal oxides	67
8.1.2	Metal sulphides	68
8.1.3	Other metallic compounds	70
8.2	Smelting practice	70
8.3	Summary of Section 8	70
	Objectives and SAQs for Section 8	71
	Objectives and Table B for Block 3 Part II	**72**
	ITQ answers and comments	**74**
	SAQ answers and comments	**78**

Study Guide and Table A for Block 3 Part II

The material in the second part of Block 3 is dealt with under two main headings — Exploration and Evaluation (Sections 1—5), and Mining, Mineral Processing and Smelting (Sections 6—8), each being equivalent to one Course Unit in length. There is a natural break between the two, and you are advised to divide your study time roughly equally between them.

Exploration for and evaluation of ore deposits is in practice a geological activity, but it is controlled by economic considerations. A theme relating to the economic costs and efficiency of exploration therefore extends throughout this part of Block 3, and the necessary data are in Table 3 (p.36). You should note that in this Part we use the terms 'ore deposit' and 'mineral deposit' synonymously.

We begin in Section 1 by looking at the economic philosophy behind the design of exploration projects, using some of the concepts introduced in Block 1. We then discuss geological surveys and techniques of remote sensing (Section 2), and geochemical and geophysical prospecting (Sections 3 and 4). Table 2 (p.25) contains all the information about minerals and rocks that you will need for the ITQs and SAQs. In Section 5 we return to economics, describing the factors that must be considered before any decision can be made to begin mining or not.

In Sections 6—8 we turn to the more technological aspects connected with ore deposits, concerned with mining them, processing the resulting ore into concentrates of ore minerals and reducing these compounds to metals. We discuss the scientific principles rather than the technical details.

Section 6 deals with the economic considerations leading to the opening of mines and to the decision on whether to operate at the surface or underground. It also contains descriptions of some important mining techniques.

Section 7 first discusses the information required to choose a sequence of processes for extracting minerals from mined rock, and the factors influencing the nature and location of processing plant. Then it examines methods of breaking rocks into appropriate sizes for ore minerals to be separated from waste rock, and finally it looks at the range of physical and chemical methods of separation, and at post-separation treatments to remove water.

Section 8 discusses the stability of metal ores and outlines the chemical principles behind the smelting techniques commonly used to obtain metals from ore concentrates.

There are two television programmes containing material particularly relevant to Block 3 Part II. These are entitled *Pine Point: origin and exploration* and *Pine Point: ore to metal.*

The one audiovision sequence (AV 6) associated with Part II is called *Remote sensing,* and you should listen to this as soon as you complete Section 2.

Table A Terms and concepts with which you should be familiar before starting Block 3 Part II

Introduced in S101*	Unit and Section No.	Introduced in S101	Unit and Section No.	Introduced in S101	Unit and Section No.
acceleration due to gravity	3, 4.3	electrical potential	8, 10.3	non-polar	12, 3.4
ampere	8, 10.3	electrical potential difference	8, 10.3	oceanic crust	4, 4.6.1
anode	12, 3.2	electrode	12, 3.2	oxidation	16 and 17, 2.2
catalyst	15, 5.3	electrolysis	12, 3.2	palaeomagnetism	5, 4.2
cathode	12, 3.2	electromagnetic radiation	9, 5	polar	12, 3.4
chemical equilibrium	14, 3	endothermic	15, 2	radioactivity	10 and 11, 4.1
chemical reactivity	15, 3	exothermic	15, 2	reaction rates	15, 5.1
conductivity	12, 3.2	induced magnetism	5, 2	reduction	16 and 17, 2.2
continental drift	6 and 7, 4.3	induced polarity	5, 2	seismic waves	4, 9
cratons	6 and 7, 2.3.3	isostatic equilibrium	6 and 7, 3.4	strain	4, 3.1
current	8, 10.3	Kelvin scale	8, 8.3	stress	4, 3.1
dipole	5, 2.2	Le Chatelier's principle	14, 3.4	tesla	5, 3.2
elastic modulus	4, 3.1	magnetic field	5, 1	volt	8, 10.3
electrical charge	8, 10.3	mole	12, 2.4		

*The Open University (1979) S101 *Science: A Foundation Course,* The Open University

Introduced in S238	Block No.	Introduced in S238	Block No.	Introduced in S238	Block No.
accessory minerals	1	fossil fuels	1	porphyry copper deposits	3I
alloy	3I	gangue	3I	quotas	1
bauxite	3I	geothermal gradient	3I	reserves	1
black sands	3I	grade	1	residual deposits	3I
chemical precipitates	3I	host rocks	3I	risk capital	1
colloidal particles	2	hydrothermal deposits	3I	royalties	1
conditional resources	1	lead time	1	screening	2
confined deposits	3I	lineaments	3I	secondarily enriched deposits	3I
cut-off grade	1, 3I	magmatic segregation	3I	sedimentary basin	3I
dip	AV 4	manganese nodules	3I	strike	AV 4
dispersed deposits	3I	overburden	2	taxes	1
economies of scale	1	overburden ratio	2	trace elements	1
elasticity of supply	1	pegmatites	3I	unconsolidated deposits	2
fines	2	placer deposits	3I	veins	3I
flocculation	2	place value	1	volcanic arcs	3I
flux	2	political risk	1		

EXPLORATION AND EVALUATION

1 Exploration strategy

Study comment In this Section we review the economic parameters and the elements of actual or potential risk which must be considered before exploration for ore deposits can begin, and which control the exploration strategy. The Section builds on concepts introduced in Sections 4 and 5 of Block 1.

1.1 Exploration, finance and risk

In most cases it is the lure of profit that encourages exploration for new mineral deposits, though political considerations may sometimes be paramount. The principal economic risk in embarking on a new exploration programme — or maintaining an existing one — is that the *lead time*A between discovery and the start of mining can be as much as ten years, sometimes more. In that time, changes in technology or the economic climate may make the venture no longer profitable. Here are some of the factors that can encourage a mining company to embark on exploration.

1   Supplies of a metal may become limited, usually because of some economic or political factor. Examples are the Sudbury strike of the 1960s and the Rhodesian sanctions of the 1960s and 1970s. In the former case, world supplies of nickel were threatened, in the second, chromium became scarce. Their *elasticity of supply*A decreased: both are essential metals and at the time very few other sources were known. The price of both metals rose, and exploration efforts were increased.

2   Improved mining technology can bring deposits that were previously in the category of *conditional resources*A into the *reserves*A category (Block 1, Figure 51). A classic example is the case of dispersed *porphyry copper deposits*A, which in about 1920 became workable at a profit, because *economies of scale*A allowed costs to fall. Such deposits became important targets for exploration.

3   Developments in industrial technology may bring into use a material for which there was previously little demand and therefore only a small number of mined deposits. Such materials can therefore become inelastic in both supply and demand almost overnight, and mining companies may scramble to get on the resulting exploration bandwagon. Examples in this category include uranium in the late 1940s and early 1950s, and metals such as niobium and zirconium, for use in *alloys*A in the space age.

4   The high capital investment involved in setting up a mine, and the high costs of just keeping it open, mean that an unexpected difficulty in mining, or the depletion of reserves may cause a once profitable mine to operate far below capacity and make a loss. Because new ore deposits a short distance away could revitalize the ore processing part of the mine, there is a high incentive to explore in active mining areas. Because all the required equipment is at hand, one of the major costs of mining is removed, and mineral deposits that would be uneconomic in their own right then become ore deposits worth seeking.

Conversely, of course, exploration programmes will be curtailed, discontinued, or simply not begun, if the economic climate is not suitable. The recession of the late 1970s and early 1980s brought about the closure of metal mines world-wide, many still with ample reserves. In such circumstances, investment in new exploration programmes is not financially attractive. However, even in times of economic recession, mining companies will keep teams of exploration geologists at work. They need to ensure that if the economic climate improves, they can quickly take advantage of a greater demand and higher metal prices.

Geological factors obviously control where to search for deposits of particular metal ores, but economic and political factors will determine the size and grade of deposits considered worth following up. Clearly, in an established mining area, the minimum size of a deposit that can be profitably mined may be orders of magnitude smaller than that of one in a little-known remote area, because expensive communications networks and processing plants already exist. The levels of *royalties*A and *taxes*[A] and the imposition of *quotas*[A] are different in different countries. Moreover, some countries may be more susceptible to sudden changes in policy towards mining than others. Because each of these many factors affects profits, a particular grade and tonnage that looks attractive in one region will be unattractive in another. We may therefore define an *exploration objective*[B] as one that comprises a metal, a region, a minimum tonnage and a minimum grade. In economic terms, it boils down to a potential profit.

exploration objective

Let us consider some of the different kinds of risk inherent in an exploration programme. The most obvious one and the only one that is concerned with geological aspects is the *exploration risk*[B] itself — the risk that despite a favourable geological setting, there are no ore deposits that match up to the exploration objective.

exploration risk

Partly because mining companies have a heavy capital commitment in existing mining districts, and partly because 'to find an elephant you visit elephant country', exploration commonly takes place in well known areas. This does mean, of course, that as time goes by there may be fewer discoveries to be made. The chances of a new discovery can be estimated from statistical models based on the frequency of occurrence of known deposits of different sizes. Figure 1 is a graph of such a frequency distribution for a mining district. Two deposits of 10^8 tonnes have been discovered, three each of 10^7 and 10^5 tonnes, and no less than ten of 10^6 tonnes. The geologists have constructed a bell-shaped distribution curve (half of which is shown in Figure 1), which they believe offers a plausible picture of the probability of ore deposits still to be discovered.

> **ITQ 1** According to the curve in Figure 1, how many *more* deposits of 10^7 tonnes and 10^5 tonnes remain to be discovered?

Of course, in practice, nature does not provide deposits in neat order-of-magnitude categories as in Figure 1, but the principle is clear enough. The model also depends heavily on the assumptions that mineralizing processes obey simple statistical rules of probability, but this is by no means always the case — as you may have gathered from Part I. You may also feel that it would be possible to draw other curves through the rather sparse data points in Figure 1. Again you would be right, but that does not invalidate the principle, which has been applied with some success in some cases to raise finance for exploration. History does not relate whether the models worked, but by then it was too late, as the money had been spent. For exploration in an active mining district, all the data from earlier work will be available and can be re-evaluated in the light of any new information or ideas about the mode of occurrence and origin of ore deposits in the district.

Clearly, new exploration in an established area is a much easier proposition than in a remote and less known region. However, there are ways of quantifying the chances of making a profitable discovery in a new area. The simplest is to compare the geology with that of active mining areas. Past records of production from the active area give an estimate of the value of metals per square kilometre that *might* be present in a geologically similar, but unexplored area. For instance, an area with a predicted value of 15 000 US dollars per square kilometre *ought* to reveal 15 million US dollars worth of metal if 1 000 km² are thoroughly explored.

In new exploration areas other kinds of risk, which are not so easily quantified, may be present. *Political risk*[A] is the most important of these at the beginning of exploration. Depending on past history, there may be a chance that the government of the area involved will increase taxation, or even nationalize mining operations. In such a case it might be wiser to seek only deposits that can be mined quickly and very profitably. Mines involving long lifetimes and high capital costs would be more likely to be adversely affected by sudden political change. As you can imagine, risk is not something that can be quantified easily, but nevertheless some estimate of the 'odds' has to be made.

If the estimated chances of satisfying the objective of 100 million US dollars profit from a discovery are 1 in 1 000 at the outset of exploration, then the absolute maximum *risk capital*[A] allowed for exploration would be 100 000 US dollars. There is very little chance that such a budget would pinpoint exactly where to open a mine. What is required is that the new knowledge helps redefine the risk, so that either more funds can be risked in further exploration, or the project can assume more modest aspirations. By the same token, if initial exploration shows the original odds to be less favourable than expected, the project can be abandoned without crippling

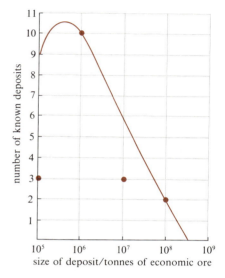

Figure 1 A frequency distribution of ore deposits containing different tonnages of economic ore.

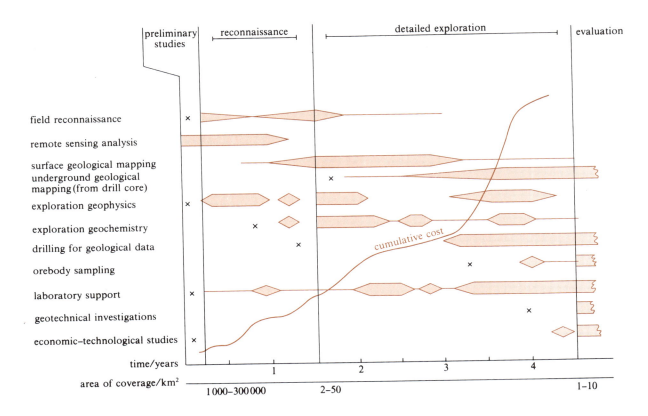

| preliminary studies | reconnaissance | detailed exploration | evaluation |

field reconnaissance
remote sensing analysis
surface geological mapping
underground geological mapping (from drill core)
exploration geophysics
exploration geochemistry
drilling for geological data
orebody sampling
laboratory support
geotechnical investigations
economic–technological studies

cumulative cost

time/years   1   2   3   4

area of coverage/km²   1 000–300 000   2–50   1–10

Figure 2 A generalized pattern of activities in an exploration programme for a poorly explored region with few existing mines or known prospects. The cumulative cost of the project gives an indication of changes at different stages in the exploration. Crosses indicate when orientation surveys for geochemical exploration are carried out, and when instruments and equipment for other surveys are evaluated.

losses. It is because of these financial requirements that exploration generally proceeds in several distinct phases (Figure 2), before each of which critical decisions on whether to continue or abandon exploration must be taken. Many of the terms in Figure 2 will be unfamiliar to you at this stage, but you will be referred back to the figure as you work through the text.

How far the ideal exploration scheme in Figure 2 is allowed to proceed and the kinds of technique that are used depend mainly on the amount of existing information about the area being explored. In an unknown and undeveloped region exploration may stop once promising areas have been identified, rather than going on to make an actual discovery. Rights to these areas may be acquired cheaply with a view to further exploration when the regional infrastructure is better developed. They may be sold later at a profit to a less adventurous but better-heeled mining company when the opportunities for opening mines are more favourable. In the case of exploration in an active mining district, all the data from earlier work will be available and can be re-evaluated in the light of what is known about actively mined deposits and new ideas about the mode of occurrence of the particular types of deposit in the district. New exploration is then initiated in much greater detail than it would be in a remote area because the odds are more favourable.

1.2 Summary of Section 1

1 The motive for exploration usually centres on potential profit, but sometimes on the need to avoid losses on existing operations. In both cases planning requires careful estimates of the future economic climate for metals, progress in mining technology and trends in global politics.

2 The exploration objective, usually expressed as a profit, consists in practice of a metal, a region, and a particular size and grade of deposit. The last two components depend largely on whether exploration is directed at an established mining district or at an area unknown in detail. They may also be conditioned by the political stability of the selected area.

3 When an exploration objective has been defined, the risk of not finding a mineable deposit must be assessed. Based on this exploration risk and the anticipated profits, a maximum budget for exploration is calculated. Exploration techniques financed by this budget aim at reducing exploration risk or indicating an abortive project as quickly and cheaply as possible. In general, the higher the political risk the greater the tendency to seek deposits with a high grade which can

give quick returns. Only relatively low risks encourage the search for deposits requiring a high investment and a large-scale long-term operation. Because of all these factors exploration proceeds as a series of more or less distinct phases (Figure 2). In practice the methods chosen in each phase depend on what is being sought, the geology of the target area, the funds available and the personal experiences of the exploration team.

Objectives and SAQs for Section 1

Now that you have completed Section 1 you should be able to:

1 Define in your own words or recognize valid definitions of the terms and concepts introduced or developed in this Section and listed in Table B.

2 Discuss the factors underlying decisions to initiate exploration, select an exploration objective, assess the risks in exploration, and assign a budget to exploration. *(ITQ 1)*

Now do the following SAQs.

SAQ 1 *(Objectives 1 and 2)*   In 1974 a politically stable country announced that it intended to promote development of a large unexplored area by special tax concessions to mineral exploration companies. What decisions would you have taken regarding exploration, assuming that the geology was favourable for the alternatives below?

(a) Would you have sought tin or copper? (cf. Block 1, Figure 15.)

(b) Would your objective have been several small deposits or one very large deposit?

(c) Would you have sought rapid returns on a high grade deposit or a long term development of a low grade deposit?

(d) Would you have assigned a high or low proportion of expected profits to exploration?

(e) Would you have planned to operate a mine as soon as possible or merely to buy rights to areas where mineralization was probable?

[NB   This SAQ is rather artificial, but sets out to apply many of the qualitative principles relating to exploration objectives.]

SAQ 2 *(Objectives 1 and 2)*   A company has decided that it could profit by finding new copper reserves. Because it wishes to explore a poorly known, remote area, the only worthwhile deposits have to contain at least one million tonnes of copper metal. Figure 3 is a plot on logarithmic axes of grade against size for different types of copper deposit. Use Figure 3 to decide what kind(s) of copper deposit must be sought. (To do this you should calculate the masses of ore with grades of 0.1, 1.0 and 10.0 per cent copper that would be needed to produce 1 million tonnes of copper, and then draw a curve on Figure 3.) If the area is in a politically insecure country is there a more favourable type of deposit?

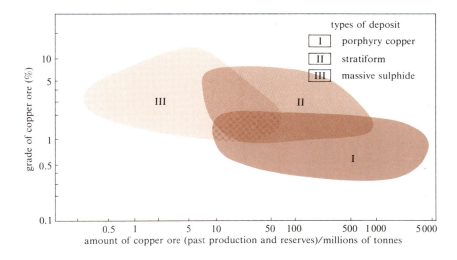

Figure 3   The grade—tonnage characteristics of the three types of copper deposit.

2 Geological exploration

Study comment This Section reviews some of the geological preliminaries to exploration and gives you an idea of the geologist's role in the search for ore deposits. We also look at modern visual aids that help the geologist work more effectively — images acquired from aircraft and satellites. You should study the audiovision sequence *Remote sensing* at the end of Section 2.3.

Until the late nineteenth century, ore deposits were discovered by a combination of luck and the 'instinct' of untrained prospectors, rather than by any systematic scientific approach. Indeed, a close look at the records of most important modern mines would reveal that they were found as a result of the persistence of a few individuals looking at rocks on the ground. However, most of the more easily discovered deposits have long since been found and many of them are worked out. With the growing demands of technological civilization, new ore deposits have in general become more difficult to locate by direct means. To compensate, however, there is now a great deal more information available about how and where ore deposits occur, as we outlined in Part I. The modern exploration geologist has geological maps at his disposal, and the availability of cheap remote sensing data from aircraft and satellites increases almost daily. These enable the modern prospector to select favourable areas for exploration, without leaving the office, and to guide field surveys in a systematic fashion.

2.1 Defining the target

After an exploration objective has been defined (Section 1.1), the next step is to select an area with geological features similar to other areas where such objectives have been achieved in the past. This means collating all existing information on the general region, in the form of maps, reports, records of earlier mining operations and so on, and using geological models for the controls of mineralization (Part I) to interpret these data. At this stage, remote sensing, geophysical and geochemical data for the region will be considered too, if they are available.

ITQ 2 In which of the settings (a)–(d) would you consider the chances of finding deposits (i)–(iv) to be highest?

(i) Porphyry copper deposits (Part I, Section 2.4); (ii) stratiform copper, lead and zinc deposits (Part I, Section 4.6.2); (iii) magmatic segregation deposits of chromium (Part I, Section 2.2); (iv) banded ironstone formations (Part I, Section 4.6.1).

(a) *Cratons*[A]; (b) ancient *oceanic crust*[A]; (c) *sedimentary basins*[A]; (d) old *volcanic arcs*[A].

When suitable *areas* for particular types of deposit have been selected, it is equally important to be sure that the appropriate rock types are sought, because it is *rock associations*[B] that determine where the geologist will go and look.

rock associations

ITQ 3 Consider the elements chromium, tin and uranium.

(a) Deposits of which of these are more likely to occur in granitic rocks and which in gabbroic rocks (cf. Part I, Table 6)?

(b) Assuming that the criteria in (a) were satisfied, would you consider prospecting for any of the three elements in (i) gravel and sand deposits in modern stream channels, (ii) sandstones with traces of plant remains (cf. Part I, Sections 4.4 and 4.5)?

It may also be useful to know about the mineralizing process itself. For example, we know that porphyry copper deposits are found in granodioritic rocks, where watery magmatic fluids have extensively altered the minerals of the host rock (Part I, Section 2.4). We do not really know why some granodiorites contain porphyry copper deposits, whereas others in similar geological settings do not. But we do know that volcanic arc settings (the site of modern or extinct subduction zones) are the best places to look for potentially mineralized granodiorites. Similarly, there is a very well known association of lead and zinc sulphide ores with limestones in the vicinity of faults (Part I, Section 3.3). We know that the ores are of hydrothermal origin, but we do not know just why some limestones are mineralized whereas others are not.

It is also worth noting that many known deposits seem to be arranged in lines or lie close to major *lineaments*ᴬ (usually large faults) at the Earth's surface (Part I, Section 5.3). This kind of pattern is another one that has been used in mineral exploration work, though it is not always wholly successful.

Sooner or later, these 'desk-top' studies and any predictions made from them must be put to the test. The geologist must go into the field, and exploration costs now begin to rise substantially (Figure 2).

2.2 Prospecting

Even in areas that are quite well known, and where probability curves of the type shown in Figure 1 can be compiled and used, the exploration geologist will probably have to revise existing geological maps. Nowadays, there are almost invariably aerial photographs or satellite pictures available, and these are a great aid. They cut down the work involved by indicating where outcrops of rock are likely to occur, but those outcrops must still be visited, observations made and measurements of *dip*ᴬ and *strike*ᴬ and fracture patterns made, so that geological boundaries can be plotted on the map.

The geological map is an essential prerequisite, because only a systematic survey will tell where the rock associations are and hence indicate likely areas for more detailed exploration. As well as showing the disposition of the rocks at the surface, the construction of cross-sections enables the underground extension of the surface geology to be determined.

In the course of this general mapping, the geologist may be fortunate enough to find surface evidence of mineralization. An obvious example is the leached capping of iron oxides that results from the oxidation of iron sulphides such as pyrite (Part I, Section 4.3), and may be an indicator of important sulphide mineralization beneath. A layered igneous intrusion may be found, in which oxide minerals may have been concentrated in bands (Part I, Section 2.2). Finds such as these *may* turn out to be ore deposits, but a great deal more work will be needed to decide whether they are.

But even if no such obvious signs of actual or potential mineralization are found by the geologist while mapping, the area may still be extremely promising. For one thing, not all the rocks will be exposed, because there is always some superficial cover of soil and vegetation and so on. For another, even the most dedicated geologist cannot visit every outcrop. Nonetheless, to continue the work in greater detail and to justify progressing to more expensive geochemical and/or geophysical methods requires that the geologist find some promising evidence of ultimate profit.

ITQ 4 Figure 4 is a vertical cross-section based on data on a geological map. It shows an igneous intrusion cutting some sedimentary rocks. Where would you concentrate a search for the following mineral deposits: (i) massive copper–nickel sulphide segregation deposits if the intrusion was of gabbroic composition; (ii) pegmatites carrying lithium, beryllium and tantalum if the intrusion was a granite; (iii) hydrothermal lead and zinc mineralization if the intrusion was a granite; (iv) secondarily enriched porphyry copper deposit if the intrusion was a granodiorite?

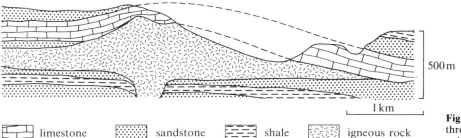

500 m

1 km

limestone   sandstone   shale   igneous rock

Figure 4 A geological cross-section through an igneous intrusion cutting sedimentary rocks.

Geological prospecting may at worst indicate that an area shows little economic potential. At best, a lone prospector might find a surface outcrop of a rich deposit, which can be mined straight away. More usually, geological prospecting gives indications of areas with potential for mineralization. To confirm that surface showings do overlie ore deposits various geochemical and geophysical techniques have also to be used. Before discussing them, we look at methods that enable geologists to acquire some data about rocks and their disposition without visiting every square metre of an area. These are the techniques of remote sensing.

2.3   Remote sensing

The Earth's surface has not yet been completely mapped geologically. Figure 5 gives an idea of this dearth of information. Since it was compiled more maps have been published, yet vast areas have only synoptic maps of geology at scales of 1 : 1 000 000 and smaller. These can help in the preliminary stages, but are of little use in the more detailed phases of exploration. Producing an accurate, detailed geological map by field work alone is a long and expensive task. On average, one person requires 100 days of field work and compilation to produce a map at 1 : 100 000 scale of an area of 100 km². This would cost 100 000 US dollars. Quite clearly, unaided geologists would need decades if not centuries to map the blank areas on Figure 5 to a useful scale. By comparison, a similar map produced by interpretation of aerial photographs would take 20 days with a cost of 10 000 US dollars. Using satellite images results in even better cost-effectiveness.

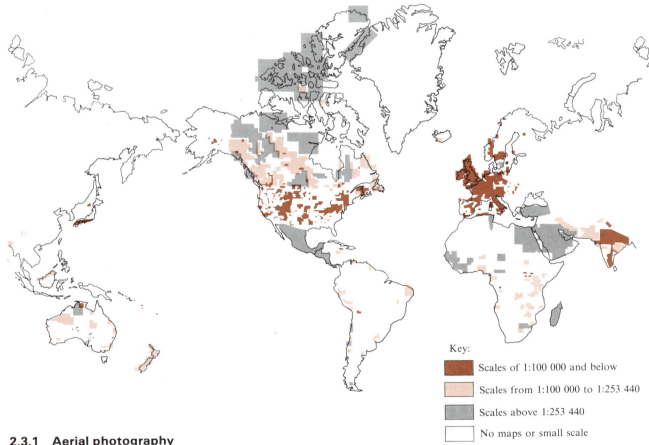

Key:

█ Scales of 1:100 000 and below

░ Scales from 1:100 000 to 1:253 440

▒ Scales above 1:253 440

☐ No maps or small scale

Figure 5 A world map showing the detail of published geological maps in 1967. There has been some improvement in North America and Australia since then. (Source: Quarterly Journal of the Geological Society of London, 1967)

2.3.1   Aerial photography

Photographs of the ground can be taken at any angle from an aircraft. However, the most useful are those looking vertically downwards, because they give a view that approximates a map, in which all features are at a similar scale. In many forested areas, a geologist can rarely see more than the ground beneath his feet. The great advantage of remote sensing is that, no matter what the terrain, details of a large area can be seen, provided they are big enough to show up. Just what can be seen depends on the resolution, which in turn depends on the grain size of the film, the flying height and the focal length of the camera. Because aerial photographs cover a great area, the continuation of surface features can be traced quite easily. Their interpretation, however, depends on the experience of the geologist involved.

Granitic rocks are light-coloured and gabbroic ones dark. Sediments and metamorphic rocks show layering and we can tell if a rock or structure is younger than the layering if it cuts or disturbs it.

ITQ 5   Look at Figure 6, in which layers in sediments are quite clear in the darker rocks of the bottom half. (i) Can you see any boundaries that cut the layering? (ii) Mark on Figure 6 the boundaries that separate areas of different grey tone. You should be able to separate three distinct tones. (iii) What is the relative age of the obvious linear features that trend roughly NNW–SSE?

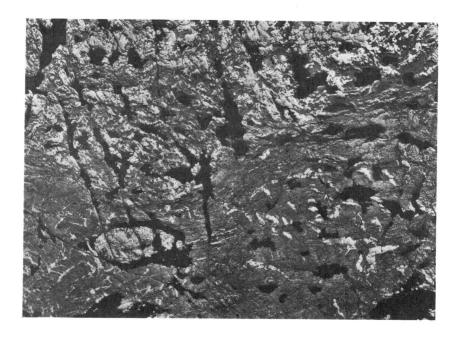

Figure 6   Vertical aerial photograph of part of northern Canada (scale 1:50 000, i.e. the area is about 5.5 km across). North is to the top.

The figure in the answer to ITQ 5 is an example of a photogeological map. It took a few minutes to compile and gives the geologist an idea of the rocks that may be found on the ground. Such deductions have to be checked by field work, though only a few traverses should suffice. Moreover, the photograph is a more accurate guide to location than any map.

Aerial photographs have their limitations, though. They are still expensive, they make use only of visible and near-visible infrared wavelengths, and cannot show details beneath cloud or dense vegetation. Much cheaper and more versatile remotely sensed pictures are now acquired using sophisticated instruments aboard aircraft and on satellites.

2.3.2 Multispectral remote sensing

Of the *electromagnetic radiation*A emitted by the Sun, some wavelengths are absorbed by the atmospheric gases and never reach the Earth's surface. Some of the radiation that does get through is reflected by surface materials, some is absorbed. The amount of radiation reflected or absorbed by a particular material varies with the wavelength. Figure 7 shows the ratio of the reflected radiation energy to the total incident energy, defined as *reflectance*B, plotted against wavelength for some plants and for some rocks. If you look at the curves for plants (Figure 7a) you can see that they reflect more radiation in the green (500–600 nm) part of the visible spectrum, than in the red (600–700 nm) or in the blue (400–500 nm). That is why grass appears green! But the most interesting feature is that all plants reflect far more infrared than visible light. This is due to the absorption of visible light, particularly red, by chlorophyll, and the strong reflection of infrared by the cell structures of healthy vegetation. If we used infrared-sensitive film to detect reflected infrared radiation, the resulting black and white image would show vegetation as bright grey patches. Another interesting feature in Figure 7a is that the contrast in reflectance between the three vegetation types is much larger in the infrared region, and in an image of reflected infrared there would be little difficulty in separating the types as different shades of grey. In images of visible bands they would all be dark grey. Figure 7b shows reflectance curves for five common rocks.

reflectance

13

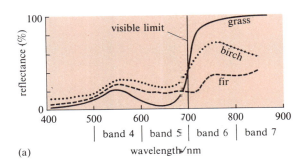

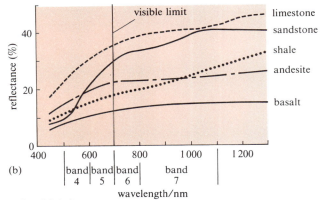

Figure 7 Spectral reflectance curves for (a) vegetation, (b) rocks. The wavebands labelled 4, 5, 6 and 7 are those from which reflectance data are gathered by a multispectral scanner aboard the Landsat series of satellites.

They are clearly different too. The sandstone would be brighter in an infrared image than the andesite, but darker in a green image.

One way of highlighting the differences in the reflectance curves of materials is to display not just one image of one waveband in tones of grey, but to combine three wavebands: one band displayed in red, one in green and one in blue. Red, green and blue are the primary light colours from which all other colours can be constructed. Because the human eye is much more sensitive to different colours than it is to different shades of grey, such a colour image is easier to interpret. The most commonly used combination is infrared, red and green reflectance displayed as red, green and blue. This produces a *false-colour image*[B]. True colour is never satisfactory from high altitudes because blue light is scattered by the atmosphere to give haze that obscures surface details. Most remote sensing systems therefore do not gather blue reflectance data. In a false-colour image grass and broadleaf trees would show up as bright red tinged with blue, conifers as dark reds and purples. The sandstone would have roughly equal proportions of red and green but low blue, and this gives yellow or brown. Limestone is a very good reflector of all wavelengths and would appear yellowish white, whereas basalt is a poor reflector and would show as dark brown. (The audiovision sequence (AV 6) uses some examples of false-colour images and explains their interpretation.)

false-colour image

There are two ways of producing false colour images. The simplest method is to use film with different layers sensitive to infrared, red and green, which when developed produce combinations of red, green and blue. The best method is to use photoelectric sensors with filters that pass only narrow bands of wavelengths. Reflected radiation is gathered by a mirror which scans narrow strips of the ground as a series of small segments. As the aircraft or satellite moves an image is built up of these regular lines and segments. The response of each sensor for each segment is converted into a number. The data are then a grid-like array of numbers, each of which corresponds to the reflectance of a small rectangle on the ground for a narrow waveband. This method has two advantages. First, being in digital form the data can be transmitted by radio to the ground, removing the need for film to be returned. Second, the digital form of the data is suitable for all kinds of computer manipulations, such as modifying contrast, automatically searching for areas with the same reflectance properties and combining data from several bands in several ways to emphasize differences.

Look at Figure 7, where you will see that the shapes of the reflectance curves differ from one material to another. These shapes can be expressed as arithmetic

Table 1 Sums of, differences between, and ratios of reflectance values in different wavebands from Figure 7

Band combination	Grass	Fir	Percentage difference	Andesite	Shale	Percentage difference
4 + 7	120	65	45	37	34	
7 − 4	80	15		10	12	
7:4	5	2		1.3	2	
7:5	31	2		1.1	1.2	
4:7	0.2	0.3		0.1	0.08	

combinations of the reflectance values at different wavelengths. We have done this for grass and fir trees, and for andesite and shale, in Table 1.

> **ITQ 6** Calculate the percentage difference between the values of the combinations in Table 1 for the two pairs of materials, and decide which combinations would be best for discriminating between the members of each pair.
>
> (Percentage difference $= \dfrac{\text{difference}}{\text{larger value}} \times 100$; we have done the first one for you.)

The answers to ITQ 6 show that multispectral remote sensing can provide the geologist with much more flexible information than that in photographs. Because of this, geologists can use their knowledge to programme a computer to find areas with particular spectral 'signatures'. In the audiovision sequence we look at means of locating zones of mineral alteration related to porphyry copper deposits as well as expanding our treatment of remote sensing.

As well as scanning reflected solar radiation, remote sensing can monitor long-wavelength infrared radiation emitted by geothermal phenomena (Block 5, *Energy Resources*) or by rocks that absorb solar radiation and eventually re-emit it at longer wavelength. In the latter case the data tell us something about the thermal properties of rocks, which differ quite markedly. Such thermal surveys are generally conducted at night so that reflected solar infrared does not swamp the sensors. Another method uses artificially produced long-wavelength radiation, radar, to illuminate the surface. As well as showing details of topography, radar remote sensing can penetrate clouds and dense jungle to give all-weather surveillance. Both these techniques are beyond the scope of this Course.

The geologist's only useful sense is that of sight. Multispectral remote sensing enlarges the range of wavelengths over which visual interpretation can be used, as well as aiding it by computer methods. Because the data can be acquired from satellites huge areas can be 'taken in' at a glance, and the cost effectiveness of preliminary exploration is improved still further. Our $100\,\text{km}^2$ area (Section 2.3) could be mapped in 5 days at a cost of $2\,000$ US dollars — 2 per cent of the cost and 5 per cent of the time for conventional reconnaissance mapping. The map would not contain first-hand geological information but would show far more detail, which could be checked by a brief field survey.

Now listen to the audiovision sequence, Remote sensing.

2.4 Summary of Section 2

1 Geological exploration begins by seeking to narrow down the area of field investigation required to satisfy the exploration objective. This is done using existing data, remote sensing and theories about how ore deposits form and with what rocks and geological structures they are commonly associated. From this, areas of high potential are identified, but equally important, areas of little potential are rejected.

2 Geological exploration in a target area aims to provide positive evidence that the exploration objective may indeed be satisfied. That is, the exploration risk must be cut down. This is achieved by finding surface indications of mineralization such as leached cappings and unusual oxidized compounds of metals or suitable geological relationships. Geological maps play an important role in recording this information, and in enabling both the evolution of the area and the extension of rocks beneath the surface to be estimated.

3 Remote sensing now plays an important role at all stages in geological exploration. Images cover much larger areas than the field geologist can encompass, so enabling field measurements to be extrapolated to unvisited areas. Multispectral techniques quantify the radiation reflected from the surface as digital numbers corresponding to the brightness of small regularly arranged segments of the surface. Consequently, these data can be both analysed by computer and displayed as easily interpreted colour images.

Objectives and SAQs for Section 2

Now that you have completed Section 2 you should be able to:

1 Define in your own words or recognize valid definitions of the terms and concepts introduced or developed in this Section and listed in Table B.

3 Use information about different rock types, their regional setting and relationships to one another to suggest sites of potential mineralization. *(ITQs 2, 3 and 4)*

4 Recognize and delineate simple geological features, such as faults, igneous intrusions and banded rocks, on remotely sensed images. *(ITQ 5)*

5 Select combinations of multispectral data that most clearly distinguish different surface materials. *(ITQ 6)*

Now do the following SAQs.

SAQ 3 *(Objectives 1 and 3)* Examine the figure in the answer to ITQ 5. Where would you concentrate a search for (a) lithium-bearing *pegmatites*[A], (b) lead–zinc *hydrothermal deposits*[A]?

SAQ 4 *(Objectives 1 and 4)* Examine colour plate 3.20. (i) What is the age of the prominent, roughly east–west and nearly straight dark lines in the western part of the image, relative to the set of curved linear features running roughly north–south? (ii) On what evidence do you base your conclusion? (iii) What do you think these two sets of features might be?

SAQ 5 *(Objectives 1 and 5)* On the basis of the reflectance curves shown in Figure 7, what would you expect to be the main differences in appearance between andesites and sandstones on images produced by printing band 7 in red, band 5 in green and band 4 in blue?

3 Geochemical exploration

Study comment In this Section we start with the definition of a geochemical anomaly and how it can be identified. We then look at the different processes whereby anomalies are produced. Different kinds of geochemical survey are used, depending on these processes. Finally we show how geochemical exploration is organized to give the best chance of locating an ore deposit quickly and cheaply.

Ore deposits are concentrations of elements far greater than those in common rocks. The rocks, soils, water or vegetation in the vicinity of ore deposits are also commonly enriched in those elements, as a result of a great variety of processes. Geochemical exploration involves the systematic search for unusual or anomalous concentrations of elements in these surface materials, which may be related to the presence of nearby ore deposits. Remember that we are dealing with metallic elements, most of which are present as *trace elements*[A] in common rocks, which means that they must be detected and measured at the parts per million (p.p.m.) level.

In exploration terms, a *geochemical anomaly*[B] is a concentration of elements that deviates significantly from the expected range of values. This expected range is known as the *background*[B], which for any element will vary from place to place and according to whether rock, soil, water or vegetation is being analysed. The boundary between background values and anomalous values is known as the *threshold*[B]. The magnitude of the anomaly (the ratio of the highest anomalous value to the average background) is termed the *contrast*[B]. You can see how this terminology can be applied to a simple anomaly by attempting ITQ 7.

geochemical anomaly

background

threshold
contrast

16

ITQ 7 Figure 8 shows some analyses of copper in soils sampled in a single traverse across a mineralized area. Look at this figure and answer the following questions.

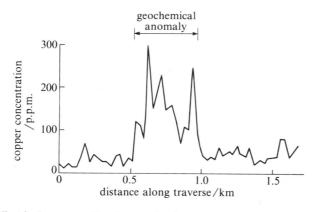

Figure 8 A copper anomaly in soil.

(a) What is the range of background values for the copper concentration?

(b) What is the range of anomalous values?

(c) What is the contrast between the anomaly and the background?

(d) What is the threshold value?

However, few exploration geochemists are lucky enough to encounter such a clear-cut subdivision between background and anomalous values. Figure 9 illustrates a more common problem. In this example, soils have been sampled at regular intervals over a 100 km² area in a reconnaissance survey and analysed for copper. The results have been plotted on to the map at each sample location. In order to interpret these data, it is very important that the correct threshold is chosen.

40	27	28	34	38	19	46	60	60	54
23	32	21	43	48	33	22	45	35	30
33	74	44	34	14	32	22	28	17	18
141	84	19	38	39	21	27	28	27	129
180	126	41	21	36	23	36	46	301	29
99	29	45	23	32	31	194	270	37	37
20	31	38	46	45	151	690	166	46	26
31	29	44	33	32	355	232	30	24	46
20	34	30	30	115	218	25	18	24	25
17	36	26	110	18	45	36	29	22	24

⊢ 1 km ⊣

Figure 9 Evenly spaced analyses for copper in soil from a reconnaissance survey.

ITQ 8 If, in the example in Figure 9, the exploration geologist selects a value of 45 p.p.m. copper in soil as the threshold, how will the pattern of further exploration differ from that suggested by a threshold of 250 p.p.m. copper? (You will need to outline the areas with values greater than 45 and 250 p.p.m. copper.)

Even supposing that an interesting anomaly is found, what guarantee is there that it lies above an ore deposit? Assessing this probability is of crucial importance because to explore further may mean using ground-based geophysical methods or even drilling, which are much more expensive. To understand the relationship between a surface geochemical anomaly and its source in an ore deposit, it is important to understand how anomalies are formed.

17

3.1 Geochemical dispersion

By now it should be clear that an ore deposit is like a needle in a haystack: very small, but very anomalous. Various geological processes, however, aid us by spreading the high concentrations of the elements related to the deposit into much larger volumes of rock or surface materials with lower but still anomalous concentrations. This process is called *geochemical dispersion*[B]. The larger 'diluted' anomalies are much easier to find than the extreme anomalies formed by ore deposits themselves, simply because they cover a larger area and the chance of sampling a dispersed anomaly is higher.

geochemical dispersion

Geochemical dispersion may have occurred during the process which formed an ore deposit itself, when it is termed *primary dispersion*[B]. If it occurred after the ore deposit formed it is called *secondary dispersion*[B]. In both cases, the dispersion of elements can take place at the Earth's surface or below it.

primary and secondary dispersion

Primary dispersion is related to the process of mineralization itself, and to understand it you will need to recall or refer back to Part I.

> **ITQ 9** For each of the ore deposits (a)–(d) briefly describe the processes controlling primary dispersion. Which deposit should have the broadest primary dispersion?
>
> (a) A magmatic segregation deposit of nickel sulphides (Part I, Section 2.2.2).
>
> (b) A porphyry copper deposit (Part I, Section 2.4).
>
> (c) A lithium-bearing pegmatite (Part I, Section 2.3).
>
> (d) A placer deposit of tin (Part I, Section 4.4).

One means of detecting anomalies produced by primary dispersion relies on the analysis of outcrops of rock, which is known as *lithogeochemical surveying*[B]. During reconnaissance, such a survey may identify rock types that favour a more detailed search, such as chromium-rich peridotite rocks, or sulphur-rich sediments which may contain copper, lead and zinc mineralization. A detailed lithogeochemical survey may delineate the broad primary metal 'halo' associated with circulation of fluids in a porphyry copper deposit (Part I, Figure 4).

lithogeochemical surveying

Many deposits however are blanketed by superficial deposits, or have narrow primary dispersion haloes (Part I, Figure 6). In these cases we have to seek the broader, weaker anomalies produced by secondary dispersion processes. This means analysing surface materials such as soils and stream sediments, spring, well and stream waters, and vegetation.

At or near to the Earth's surface an ore deposit is subject to chemical and physical weathering, erosion and transportation of the weathering products. The metal in the deposit can be dispersed in two possible forms; either as solid particles of the ore mineral itself, or in solution as ions or *colloidal particles*[A]. Solid particles may be either left in place as parts of soils or transported by gravity, water, wind or ice, and thereby removed from the area or laid down near to the deposit. Metals dissolved in water will enter either surface drainage or subsurface water systems. In these forms they may be (i) transported away from the deposit, (ii) taken up by vegetation as part of the life cycle, (iii) precipitated in stream sediments or in soils as a result of various chemical reactions.

There are therefore many secondary dispersion processes, each producing a different type of secondary geochemical anomaly. You have encountered many of the factors involved already (Part I, Section 4.1). To recap briefly, properties of the transporting medium, such as its energy, viscosity, and ability to dissolve metals, play important roles in the dispersion of metals. So too do topographic features, such as slope angle. The hardness, density and solubility of ore minerals also determine how far metals can be dispersed. We shall deal with only a few materials that illustrate different secondary dispersion patterns: soils, stream waters, stream sediments, and vegetation.

3.2 Geochemical surveys of soils

Except in the case of constantly moving sediment, such as sands in deserts, river beds and beaches, and that of deposits transported by glaciers, the unconsolidated

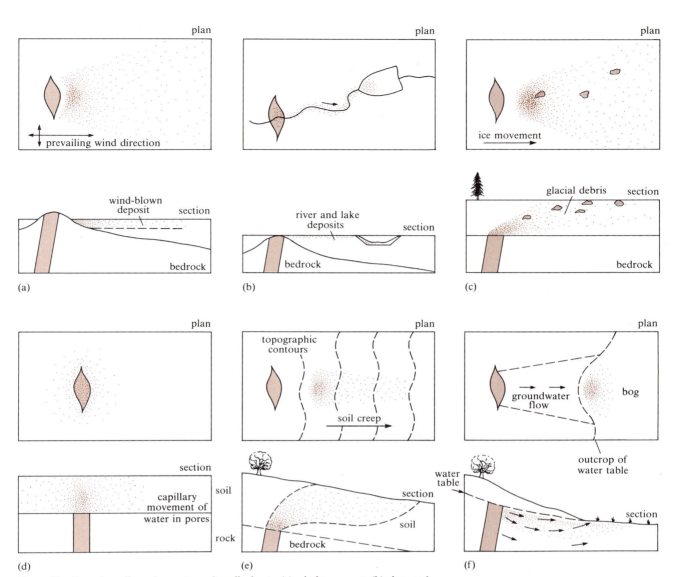

Figure 10 Secondary dispersion patterns in soils due to (a) wind transport; (b) river and lake sedimentation; (c) glacial transport; (d) movement of soil water; (e) soil creep; (f) groundwater flow. Note that only in (a) and (b) does the deposit form an outcrop.

material or soil lying on top of bedrock is more or less stabilized over periods of thousands of years. It achieves a geochemical balance between the rock from which it was derived and the processes of rock weathering, movement of water in pore spaces, and biological activity. Soil frequently has a layered structure resulting from this balance, and in a *soil survey*^B the layer giving the clearest contrast between anomalies and background is selected for analysis. Once the data have been gathered, the problem is how to use any anomalies to locate ore deposits.

soil survey

> Can you describe the sort of terrain in which a geochemical anomaly in soil lies directly above a geochemical anomaly?

Firstly the ground has to be flat, and second the soil must result solely from the downward 'rotting' of rock. In most cases these conditions are not met perfectly, and anomalies are shifted away from the deposit. Even on a flat surface the soil may comprise particles that are being moved along or have been transported before being stabilized as a soil. In the first category are wind-blown sands (Figure 10a), and in the second are sediments in a river valley (Figure 10b) and those deposited from glacial ice (Figure 10c). In the absence of sediment transport, the anomaly could still be dispersed by movement of dissolved metals in soil water (Figure 10d). The presence of a sloping surface further complicates matters. Soil particles gradually move downslope, or creep, through the action of gravity, resulting in displacement of any anomaly (Figure 10e). Groundwater that penetrates the ore deposit will flow downslope and may produce soil anomalies where it percolates to the surface (Figure 10f).

3.3 Geochemical surveys of drainage

The dispersion patterns in stream water and stream sediments, which are examined in *natural water*[B] and *drainage sediment surveys*[B], result from another set of intricate mechanisms. Where a stream has not crossed an ore deposit, you would expect to find low metal values in both water and sediment. When a stream crosses a deposit then high values in one or both appear as soon as the deposit is crossed. Downstream of the deposit there will be, to a greater or lesser degree, anomalously high metal values as a result of transport from the deposit. In a survey the analyst works upstream and follows tributaries which are anomalous, until the highest values suddenly stop at a *cut-off point*[B]. This is the point in the stream closest to the primary anomaly and, hopefully, to an ore deposit. The distance over which the anomaly extends downstream before merging with the background is known as its *persistence*[B]. Before examining this parameter in a little more detail we need to establish under what conditions any anomaly will be produced. First, this depends on the nature of the ore minerals in a deposit. A natural water anomaly will result only if an ore mineral is soluble. Insoluble minerals may contribute to drainage sediment anomalies, but only if several conditions are satisfied.

natural water survey
drainage sediment survey

cut-off point

persistence

> **ITQ 10** Which of the following minerals is most likely to contribute to drainage sediment anomalies? (i) A low density, soft mineral; (ii) a high density, soft mineral; (iii) a hard, high density mineral; (iv) a hard, low density mineral.

There are other factors that further complicate drainage anomalies, for instance water chemistry. One mineral may be dissolved in acid water, another in more alkaline conditions.

The persistence of an anomaly is governed by two main factors. The first is the degree to which it is diluted by water and sediment entering the stream from tributaries that have flowed over barren ground, or by springs and barren sediments in the channel of the stream itself. Second, changes in water chemistry may cause dissolved ions to be precipitated on the stream bed, and in analogous fashion a decrease in stream energy may deposit all the dense mineral particles, thereby stopping the anomaly spreading downstream. You should note that precipitation of anomalous ions from water on to sediment grains will increase the anomaly in stream sediments, but decrease it in stream water.

> **ITQ 11** Figure 11 shows three stream systems, each of which flows over an identical mineral deposit. Which system will display the greatest persistence?

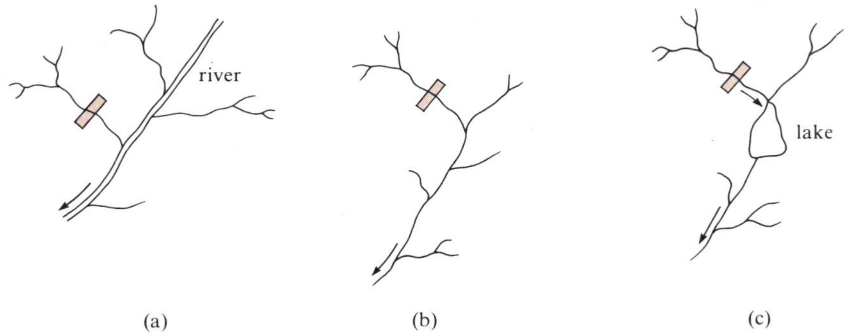

(a)　　　　　　　　　(b)　　　　　　　　　(c)

Figure 11 Different stream patterns flowing over a mineral deposit.

3.4 Vegetation surveys

Most plants can tolerate various ranges of concentration of different elements in soil and groundwater. At one extreme there is a minimum concentration of a vital nutrient such as phosphorus or potassium, at the other a concentration of an element above which the element becomes toxic to the plant. In some extreme cases of metal-rich soils only very hardy or specialized plants can survive. Such a case is reflected by a barren patch lacking vegetation or containing a few species of metal-tolerant plants. Some of these plants are very distinctive and can be used in *geobotanical surveys*[B] as *indicator plants*[B]. Zinc-rich soils in Central European mining districts often have thriving communities of calamine violet with yellow flowers that were used as a prospecting guide in the Middle Ages. Another means of

Colour plate 3.25
geobotanical survey
indicator plants

detecting anomalies is based on stunting and discoloration — usually yellowing — of leaves on plants growing in metal-rich soils. Such changed foliage indicates *stressed vegetation*[B]. These visible changes in the vegetation cover of an area can be sought by remote sensing, particularly when they involve a colour change. In that case multispectral techniques can be powerful exploration tools.

Biogeochemical surveys[B] use chemical analyses of foliage as a means of detecting anomalies in soil and groundwater. They are no cheaper than soil surveys, and suffer from the drawback that leaves and twigs have first to be dried or burnt in an oven before comparable analyses can be made. Moreover, the relationship between the metal content of a plant and the soil in which it grows is often complex. A cautionary tale concerns the Arctic dwarf birch, whose metabolism concentrates zinc, even on ordinary soils, so that its ash may contain up to 1 per cent zinc. This is greater than that of most other plants growing directly upon a zinc orebody. However, an important advantage of biogeochemical surveying is that the roots of trees may penetrate very deep soil and bring up anomalous metal contents which would otherwise remain undetected (Figure 12).

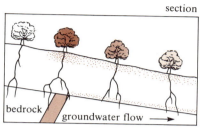

Figure 12 The biological dispersion of an element in leaves and soils (through leaf fall) from a mineral deposit covered by thick soil. This is typical of tropical rain forest.

3.5 Planning a geochemical survey

Compared with any other type of exploration, much more topographic and geological information about the terrain to be explored needs to be known when planning a geochemical survey, because the topography and to a lesser extent the geology determine the way in which target metals have been dispersed. In the case of exploration in a new area, therefore, geochemical surveys only begin in earnest after a thorough reconnaissance using highly cost-effective remote sensing, airborne geophysical surveys (Section 4) and rapid field geological surveys. Geochemical surveys are relatively slow, because samples have to be collected and analysed carefully, and such surveys are more expensive than simple geological surveys. Rough estimates of the areas covered per day and the costs per square kilometre are 25 km² and 50 US dollars for drainage surveys and 2 km² and 750 US dollars for both soil and biogeochemical surveys. As in all types of survey, geochemical exploration proceeds in a series of stages. In general, reconnaissance is based on drainage surveys whereas detailed exploration focuses on soils and, where appropriate, vegetation. However, some strategy has to be planned at the very earliest stage.

Initially an *orientation survey*[B] is carried out (Figure 2). This considers all the available data on climate, stream patterns, soil types, vegetation cover and recent geological history which may have imposed special types of dispersion such as that produced by glaciation or desert erosion. It establishes the best material to be sampled, that is whether bedrock, soil, vegetation, drainage water or sediment, or a combination of several, will show the greatest contrast between anomalies and background.

Because most of the analytical techniques need to be simplified for direct use in the field, they are often designed to show *estimates* of metal concentrations, such as between 500 and 1000 parts per million. To get precise analyses is time-consuming and requires complex, heavy equipment. Consequently, it is vitally important to decide on a threshold value between anomalies worth following up and those that can be ignored. This was demonstrated by ITQ 8. We return to the example used there as an illustration of this part of the reconnaissance survey.

> **ITQ 12** There are 100 data points on Figure 9 covering 100 km². (i) For each point, decide in which histogram division in Figure 13 the soil copper content falls and mark it as an addition to that histogram bar. Where a value falls on a division, e.g. 25 or 77, mark it in the bar to the right. *This has already been done for the top four rows of samples.* (N.B. The log scale for concentration in Figure 13 compresses the data and gives equal weight to all analyses.)) (ii) Draw a smooth curve joining the tops of each histogram box. (iii) What does the histogram tell you about the distribution of soil copper content in the area? Identify background and anomalous samples. (iv) What concentration of copper should be chosen as the exploration threshold? (v) Use the value from (iv) to shade in possibly significant anomalies on Figure 9.

The anomalies you discovered in ITQ 12 do not necessarily lie over ore deposits. They may have been displaced during secondary dispersion, or no rich material may be present. However, you now have a refined target for further work. The next step is concerned with how to zero in on a possible orebody or reject further areas from

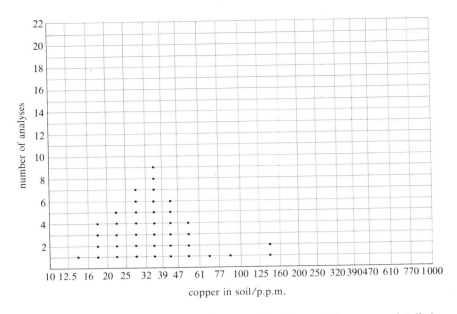

Figure 13 Log–linear graph of sample concentration against frequency of occurrence (to be completed as a histogram with data from Figure 9).

consideration in the most cost-effective way. Say the next stage was a detailed drainage sediment survey (Figure 14) in the broad area of soil anomalies indicated by the reconnaissance survey of Figure 9. The streams shown on Figure 14 are draining one of the anomalous areas identified on Figure 9. The drainage survey would progress up the main stream, analyses being made near the mouth of each

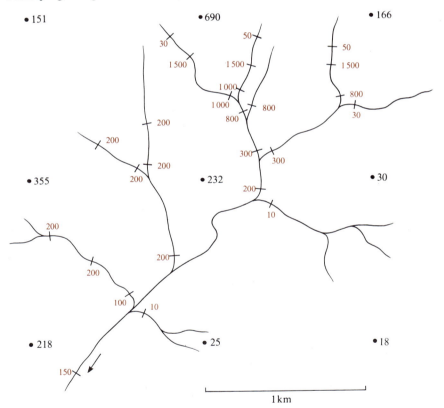

Figure 14 An enlargement of part of Figure 9, showing streams near the anomaly detected by the reconnaissance survey. Copper values (brown) in the stream sediment are shown at the minimum number of localities necessary for cut-offs to be detected.

tributary. Those tributaries showing low copper content in sediment would be rejected, those showing high values would be sampled systematically to their sources. As it turns out, the northernmost tributaries have the most anomalous sediments, and only they show cut-off points, at values of 1 500 p.p.m. copper (Figure 15a). These cut-offs enable the approximate trend and extent of copper-rich rocks to be identified.

At this stage the strategy must change. Only samples of soil or vegetation from between the streams can improve the precision with which the area of interest is

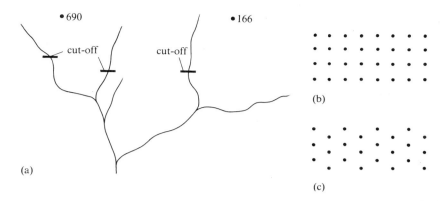

Figure 15 (a) A part of Figure 14, showing the cut-offs from a detailed drainage sediment survey. (b) and (c) Two possible sampling patterns for a detailed soil survey in the area near the cut-offs.

explored. Rather than haphazard sampling, a carefully designed programme must be employed.

> The two patterns shown on Figure 15b and 15c would have identical numbers of sample points. Which would be more likely to detect an anomaly trending east to west?

The pattern in Figure 15b has gaps in both east–west and north–south directions, and may miss a narrow anomaly. The pattern in Figure 15c has much narrower east–west gaps, and is the one most commonly chosen. The actual sample spacing is more difficult to decide on, and this depends on how much is known about the kind of deposit being sought. If the initial exploration objective rejected small high-grade deposits for economic reasons and favoured a large low-grade deposit, then there is no reason to have a narrow spacing.

You should note that the example we have followed from ITQ 12 to here is unusual, in that cheaper and faster drainage surveys generally precede soil surveys, whereas here we started with a widely spaced soil survey to bring out the most important aspects of exploration strategy.

Ultimately, geochemical exploration of surface materials will be complete. It now remains to interpret the data in the light of knowledge about the dispersion processes involved, the local geology and the results of geophysical surveys, which will probably have been performed at the same time (Figure 2). At best the search has been focused on a small area where all surveys agree that an ore deposit is highly likely to be present below the surface. In this case the next step is drilling for subsurface samples. More usually, things are not so certain and confirmation is sought using detailed geophysical surveys.

3.6 Summary of Section 3

1   Geochemical exploration is the search for anomalous concentrations of elements, generally in materials exposed at the Earth's surface such as rocks, soils, stream waters and sediments, and vegetation.

2   A geochemical anomaly is a significant deviation from the expected range of values, or background. For each survey a threshold concentration is chosen to distinguish background from anomalies and cut down the area of interest.

3   Each type of survey involves orientation, field work on reconnaissance or detailed scales, interpretation and co-ordination with geological and geophysical surveys.

4   Interpretation of data relies on knowledge of the various processes of dispersion that may be involved with different types of deposit and in different areas. Primary dispersion of elements occurs during mineralization and depends on the mineralizing process itself. Secondary dispersion occurs after mineralization and is caused by the transport of elements during later physical, chemical and biological processes at or near to the surface. Commonly, several kinds of secondary dispersion have been involved in producing a geochemical anomaly.

Objectives and SAQs for Section 3

Now that you have completed Section 3, you should be able to:

1 Define in your own words or recognize valid definitions of the terms and concepts introduced or developed in this Section and listed in Table B.

6 Using data on the distribution of an element over an area, suggest a threshold, identify anomalies and describe them, and suggest further sampling patterns. *(ITQs 7, 8 and 12)*

7 Outline the characteristics of different primary and secondary dispersion patterns and account for them in terms of the dispersion processes involved. *(ITQs 9 and 10)*

8 Interpret examples of data from geochemical surveys of: (a) bedrock; (b) soil; (c) natural water; (d) drainage sediments; (e) vegetation. *(ITQ 11)*

Now do the following SAQs.

SAQ 6 *(Objectives 1 and 7)* Decide whether each of the dispersion processes (i)–(iv) is primary or secondary. (i) Dispersion of boron in a granite by fluid with a temperature of 330 °C. (ii) Diffusion of metals from a massive sulphide deposit during metamorphism. (iii) Precipitation of manganese from stream water in a bog. (iv) Transport of large fragments (1 tonne) of massive sulphide by a glacier.

SAQ 7 *(Objectives 1 and 8)* Figure 10f shows a geochemical anomaly discovered in a bog. What types of geochemical survey other than analysis of sediment in the bog might be used in 'zeroing in' on the mineral deposit?

SAQ 8 *(Objectives 1 and 6)* Describe numerically the background and the contrast of the main anomaly in Figure 9, based on your answer to ITQ 12.

SAQ 9 *(Objectives 1 and 7)* Which of the following metals might be sought by a drainage sediment survey: mercury, nickel, chromium, copper, tin? (Use the data in Table 2 (Section 4); assume that the ore minerals listed there are the main ores for each metal, and compare their properties with those of common rocks.)

4 Geophysical exploration

Study comment In this Section we describe the measurement of different force fields, which can be used to detect variations in physical properties of rocks, and some electrical phenomena related to certain kinds of mineral deposit. Special attention is paid to the relationship between the shapes of anomalies and the shape, attitude and depth of mineral deposits.

As well as their visible and chemical properties, ore minerals, mineral deposits and the rock types that commonly contain them also possess a great range of physical properties (Table 2). These include: the *elastic modulus*[A], which partly determines the velocity of *seismic waves*[A] passing through them; density; thermal conductivity; electrical resistance; magnetic susceptibility; and *radioactivity*[A] based on the content of naturally occurring unstable isotopes.

The only way of evaluating some of these physical properties, such as thermal conductivity, is from rock samples in a laboratory. Because of this they cannot be exploited directly by any exploration technique: they are variables in very complex processes, such as the role of thermal conductivity in producing anomalies in the Earth's heat flow (Block 5). Other properties cause variations in natural force fields, such as the effect of density on the gravitational field. Measurements of these fields will show *geophysical anomalies*[B] of various kinds. Electrical properties of rocks, such as resistance, can be measured or detected by putting energy in the form of electrical currents into the ground and looking for variations in *electrical potential difference*[A] at the surface. In a similar fashion, artificial seismic waves

geophysical anomalies

Table 2 Chemical and physical properties of some common ore minerals and rocks

Metal	Mineral or rock	Hardness, Moh's scale*	Solubility	Average density/ kg m⁻³	Magnetic susceptibility/ × 10⁻⁹ SI units	Resistivity/ ohm m
lead	PbS (galena)	2.5	$10^{-5.8}$	7 600	0.2	1.9×10^{-2}
titanium	FeTiO₃ (ilmenite)	5.5	10^{-12}	4 700	1 880	4.0
nickel	(Fe,Ni)₉S₈ (pentlandite)	3.5	1	5 000	450	1.1×10^{-5}
copper	CuFeS₂ (chalcopyrite)	3.5	1	4 200	0.4	1.4×10^{-3}
mercury	HgS (cinnabar)	2.5	$10^{-3.9}$	8 100	0.2	1.0×10^{-3}
gold	Au (native gold)	2.5	10^{-18}	19 300	0.2	2.4×10^{-8}
iron	Fe₃O₄ (magnetite)	6	10^{-12}	5 180	6 280	0.6
iron	FeS₂ (pyrite)	6	10^{-7}	5 000	1.6	3×10^{-2}
iron	FeS (pyrrhotite)	3.5	10^{-7}	4 650	1 571	1.0×10^{-4}
tin	SnO₂ (cassiterite)	6.5	$10^{-6.3}$	6 920	1.1	0.2
zinc	ZnS (sphalerite)	3.5	1	3 750	0.8	1.5
chromium	FeCr₂O₄ (chromite)	6	10^{-10}	4 600	8	1×10^3
	granite	6	—	2 640	3	3×10^2 to 10^{10}
	gabbro	6	—	3 030	75	10^3 to 10^6
	shale	3	—	2 400	0.6	10 (wet)
	limestone	4	—	2 700	0.3	0.6×10^3 (wet)
	sandstone	6	—	2 350	0.4	4×10^3 to 10^8

enable us to probe the elastic properties of rock at depth by measurement of the time it takes for them to pass through rock and be reflected or refracted back to the surface. Since the seismic technique is mainly used in exploration for *fossil fuels*[A] we deal with it in Block 5. Perhaps the simplest property to survey is that of radioactivity. If a rock has a high content of radioactive isotopes of potassium, uranium and thorium, the radiation they emit at the surface and into the air can be measured directly by radiation counters. Because the type of radiation is governed by the isotope that emits it and its intensity is proportional to the concentration of the isotope at or near to the Earth's surface, *radioactivity surveys*[B] are really an indirect means of geochemical surveying.

*Hardness values for rocks are for their main minerals.

radioactivity survey

Each property can be exploited by several survey techniques for exploration purposes. Most of them are complex in terms of the instruments used and the mathematical methods involved in data interpretation. For these reasons we cover in this Course only the most basic of principles of gravity, magnetic and electrical survey techniques.

Geophysical surveys differ from geochemical surveys in an important way: some of them can be conducted from the air as well as on the ground. There are indeed means of detecting large magnetic and gravity anomalies from satellites. Geophysical exploration therefore offers a range of cost, efficiency and sensitivity that can be exploited to make the most cost-effective use of funds, equipment and personnel. As an example an airborne survey for reconnaissance can cover 500 kilometres per day in a pattern of flight lines at a cost of 25 US dollars per kilometre, whereas a detailed ground survey can cover 10 kilometres per day at a cost of 150 US dollars per kilometre. Because of this, 'Exploration geophysics' on Figure 2 includes early airborne surveys and several ground surveys at later stages.

Like geochemical surveys, geophysical surveys are frequently conducted according to a careful plan aiming at efficient and systematic data collection. More often than not data are gathered at equally spaced points along a carefully surveyed grid of straight lines. These lines may be followed by an aircraft or physically constructed on the ground as roadways.

4.1 Gravity surveys

The variation in density between different rock units in the Earth gives rise to disturbances of the Earth's gravitational field, expressed by the *acceleration due to gravity (g)*[A]. However, the density of rocks and minerals in the Earth's crust varies only in the range from 2 000 to 20 000 kg m⁻³; some other properties vary by factors of many orders of magnitude (Table 2). The acceleration due to gravity varies according to the mass of a body and its distance from the point of measurement. For bodies as large as the Earth, the gravitational field is so strong that very large masses of different density tend to sink or rise within the crust, thereby tending to equalize g over the whole surface. This is the principle of *isostatic equilibrium*[A]. However, because the strength of rocks is high, it is possible for masses which are very small relative to that of the whole Earth to be supported out of equilibrium. Such masses produce *gravity anomalies*[B] which can be detected.

Anomalies produced by bodies with a lower-than-average density in the crust, but which are prevented from 'floating' upwards by the strength of the crust, result in a lower g value than normal — a negative anomaly. Likewise, denser bodies that cannot sink produce positive anomalies. The magnitude of the anomaly is determined by two factors: the difference in mass between the anomalous body and that of a hypothetical body of similar volume composed of rocks of average crustal density; and its distance from the point of measurement. Consequently a large deep body may have the same effect as a small shallow one.

Further complicating the picture is the shape of the body, i.e. how the mass deficiency or excess is distributed in the crust. Figure 16 shows that the same shape of gravity anomaly can be produced by bodies with different volume, shape and depth. The interpretation of gravity anomalies is therefore no simple matter, and depends on mathematical models constrained by knowledge of local geology and the probable shapes of mineral deposits and other structures.

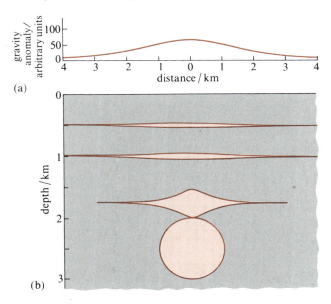

Figure 16 Four dense bodies of different shape and depth of burial (b) which would produce the same gravity anomaly measured at the Earth's surface (a). The density contrast between the bodies and surrounding rocks is $1\,000\,\mathrm{kg\,m^{-3}}$ in each case.

In Part I you saw that mineral deposits have an apparently bewildering variety of shapes, sizes and orientations. However, for the purposes of a gravity survey they can be divided into three fundamental shapes — spheres, cylinders (pipes) and sheets. Figure 17 shows model anomalies developed over anomalously dense bodies with these three shapes. In cross-section the anomalies for spheres and pipes have

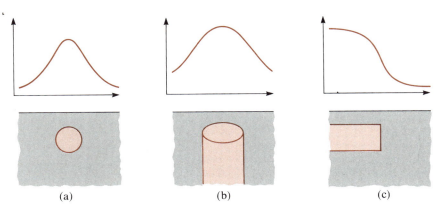

Figure 17 Gravity anomalies calculated for (a) a sphere; (b) a vertical cylinder; (c) a horizontal sheet. In each case the body is denser than its host rock by the same amount.

similar but subtly different shapes. However, the shape for a horizontal sheet is totally different, being simply a step from a positive anomaly over the sheet to normal gravity. In plan view the contours of gravity values are circular for spheres and cylinders but roughly straight lines for the edges of sheets. For bodies with anomalously low density negative gravity anomalies are produced, which are approximately mirror images of those in Figure 17.

Figure 18 shows how the shapes of gravity anomalies vary with the depth of burial and orientation of anomalously dense sheets. You should note that for a deep sheet the gravity anomaly is both lower and broader than for a shallow sheet. Another important point is that for vertical sheets the anomalies are symmetrical. For inclined sheets the anomalies become progressively asymmetrical: the peak values lie directly above the shallowest parts of the sheets and tail off in the direction towards which the sheets dip deeper into the Earth. What is not shown in Figure 18, but which should be obvious, is that the 'height' of an anomaly also depends on just how much more dense than the surrounding rock the anomalous body is. This brings us on to the question of whether or not a particular mineral deposit can be detected by gravity surveys.

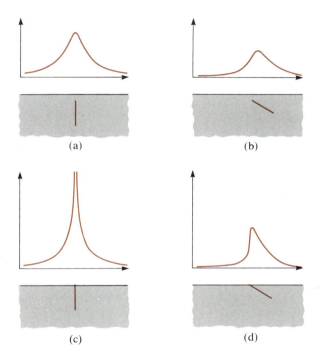

(a)

(b)

(c)

(d)

Figure 18 The variation in shape of gravity anomalies for sheets with different orientations, and different depths of burial. In (a and b) the sheets are deeply buried; in (c and d) the sheets reach the surface.

Some deposits, such as porphyry types, consist of large volumes of common rock which often contain less than 1 per cent by weight of ore minerals. Even if the ore minerals were five times more dense than the host rock (highly unlikely) their net effect on the density of the deposit as a whole would be no more than a 5 per cent increase. They would not be easily detected by gravity surveys, especially if buried. On the other hand, a very rich deposit, such as a hydrothermal vein or magmatic segregation deposit containing high-density sulphides, may have a much greater density contrast with its host rock and might thereby produce a detectable gravity anomaly. However, such deposits are quite frequently small in volume, and gravity anomalies will consequently be localized. To find them the gravity survey stations have to be closely spaced, resulting in decreased efficiency and increased cost.

You should be able to surmise from this that gravity surveys may have a role in detailed exploration for rich deposits of dense ore minerals, but only when the likelihood of discovery has been indicated by other exploration methods, such as drainage geochemical surveys, and when other geophysical methods do not work. They can play a much more useful role early in exploration by helping to define the geology beneath an area. For instance, one of the criteria for hydrothermal mineralization is that there must be a source for the metals (Part I, Section 3.1.3).

Sometimes this can be a particular type of rock, such as a granite for metals like tin and uranium or basaltic lava for copper and gold. If either is deeply buried beneath sediments, hydrothermal veins could have formed in the sediments from watery fluids rising from the source rock. If these source rocks were surrounded by metamorphic rocks (density $2\,750\,kg\,m^{-3}$) then the presence of granite (density $2\,640\,kg\,m^{-3}$) may produce a negative anomaly, whereas basaltic rocks (density $2\,900\,kg\,m^{-3}$) show up as a gravity high. This sort of information could direct exploration towards the gravity anomalies. For this reconnaissance gravity survey, the basic data may come from gravity anomaly maps compiled by government surveys.

4.2 Magnetic surveys

Unlike its gravitational field, the Earth's *magnetic field*[A] has both a variable magnitude and a variable direction. Both variables are expressed in Figure 19, where the spacing of lines of magnetic field reflects the strength of the Earth's magnetic field. Magnetic surveys depend on the Earth's magnetic field being locally distorted by *induced magnetism*[A] in rock, which in turn depends on a rock's content of minerals with a high *magnetic susceptibility*[B], or ability to become magnetized. Induced magnetism is produced in magnetically susceptible materials when they are in a magnetic field such as that of the Earth, and is responsible for iron filings being attracted towards a bar magnet. The intensity of induced magnetism (I) is proportional to the strength of the Earth's field (H) in *teslas*[A] (T):

$$I = \chi H$$

The variable χ is the material's magnetic susceptibility. The Earth's field is about 0.5×10^{-4} T, and a convenient unit for induced fields is the *nanotesla*[B] (nT) (1 nT = 10^{-9} T). Sometimes, magnetic minerals in rocks have retained a *palaeomagnetism*[A] which was induced in them by the Earth's magnetic field when they formed. Because of *continental drift*[A] and variations in the Earth's field, palaeomagnetism will have a different intensity (usually lower) and direction compared with induced magnetism. For simplicity, we ignore palaeomagnetism and just examine magnetic anomalies due to induced magnetic fields.

Only three *accessory minerals*[A] have a high enough magnetic susceptibility to produce measurable anomalies of induced magnetic field. These are magnetite (Fe_3O_4), ilmenite ($FeTiO_3$) and pyrrhotite (FeS) (iron metal does occur, but only very rarely). Consequently, magnetic anomalies reflect the proportion of these minerals in rocks. Because the iron content of common rocks varies, and all the magnetic minerals are iron compounds, magnetic anomalies indirectly reflect rock composition. Therefore, magnetic surveying, like gravity surveying, is a useful means of outlining subsurface geology. Because the magnetic minerals are frequently associated with other, non-magnetic ore minerals, such as magnetite with chromite and pyrrhotite with nickel sulphides (Part I, Section 2.2), the method can give direct indications of orebodies. You should note that other iron minerals, such as hematite (Fe_2O_3) and iron-rich silicates, are not very magnetic because of their molecular structure, so if an iron-rich rock does not contain any of the magnetic minerals it will not have much induced magnetism.

Magnetic force fields, which have variable direction as well as varying magnitude, can be measured in three ways: as the total field strength, the strength in the vertical direction or that in the horizontal direction (Figure 20). The most common and cost-effective survey method is to tow a magnetometer behind an aircraft. However, only the total field strength can be measured because of problems with stabilizing the magnetometer. Because airborne magnetic surveys are the most important we will only examine total field strength anomalies.

Figure 21 shows a magnetized body beneath the surface acting as a dipole with lines of induced magnetic field. Also shown are the lines of the Earth's magnetic field and the total magnetic field anomaly measured above. Note that the dipole is parallel to the inclination of the Earth's field, and because this inclination varies with latitude, induced dipoles therefore have variable orientation (see Figure 19). The important thing to note is the shape of the anomaly. Where the induced field is in the same direction as the Earth's field, the total field is increased. Where the fields are opposed, the total field is diminished. Thus at B both fields are exactly parallel and in the same direction, so the anomaly reaches a maximum. At D they are parallel but in opposite directions, so the anomaly is at a minimum, less than the Earth's field.

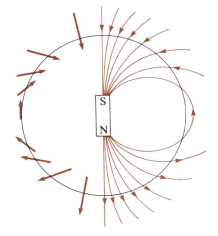

Figure 19 A simplified cross-section of the Earth's magnetic field in relation to a *dipole*[A] in the core. The Earth's north magnetic pole is a south-seeking pole, so the diagram has geographic north at the top. Both left and right parts of the figure express the strength and direction of the magnetic field and how they vary with latitude; on the left, field strength is proportional to the length of the arrows, on the right it is proportional to the density of lines of magnetic field.

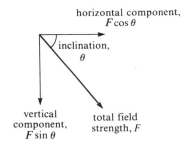

Figure 20 The relationship between the total magnetic field strength, the angle of field inclination and the vertical and horizontal components of field strength.

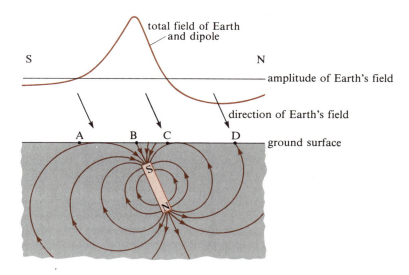

Figure 21 The relationship between the Earth's magnetic field and an induced magnetic field in a body of rock containing magnetic minerals. The induced field is represented as a dipole.

At A the fields are at right angles so the induced field has little or no effect on the Earth's total field. The further away from the dipole, the wider the spacing of the lines of the induced magnetic field, which means the induced field becomes weaker, and the anomaly decreases to zero. This also explains why the negative anomaly at D is weaker than the positive one at B.

The inclination of the Earth's magnetic field has a strong effect on the shape of anomalies.

ITQ 13 Figure 22 shows a dipole induced by the Earth's magnetic field at the equator, where the inclination is horizontal. Draw in the lines of the induced magnetic field, remembering that they are in the *same direction* as those for the Earth's field at the dipole poles; use Figure 21 as a guide to their shape and spacing. Using the same principles that we applied to Figure 21, sketch in the shape of the magnetic anomaly associated with the dipole.

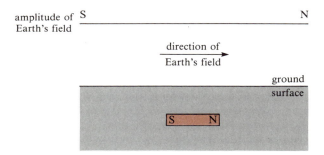

Figure 22 An induced magnetic dipole in a body of rock containing magnetic minerals buried beneath the Earth's magnetic equator.

As with gravity anomalies, the 'sharpness' of measured magnetic anomalies depends on the distance between the measuring instrument and the anomalous body. Figure 23 shows how the anomalies due to two magnetic bodies buried in the Earth become progressively less marked and broader as this distance increases, until they cannot be resolved from a single broad 'hump'. This effect is produced either by varying the flying altitude or by different depths of burial. Because crystalline igneous and metamorphic rocks most commonly contain magnetic accessory minerals, and are often a mixture of many rock types, when close to the surface they are represented by a jumble of many sharp anomalies. This gives a high *magnetic 'relief'*[B]. Deep burial by less magnetic sediments results in blurring of these anomalies until they are eventually swamped by the effect of sediments. This gives areas with low magnetic 'relief'. As a general rule, therefore, magnetically rough zones correlate with near-surface crystalline rocks, whereas smooth zones represent deep sedimentary basins. Because sediments have lower magnetic susceptibility than the crystalline rocks that form the bulk of the crust (Table 2), such basins usually show *negative* anomalies too.

magnetic relief

As with gravity surveys, those based on the detection of magnetic anomalies are very difficult to interpret in terms of the shapes of the anomalous bodies. Nevertheless, they give broad indications of the geological structure of an area. As well as

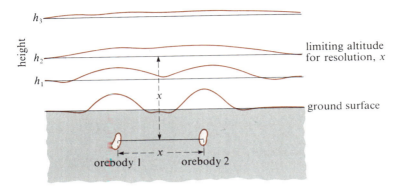

Figure 23 The anomalies due to two magnetic bodies can be distinguished by measurements close to the surface, but as the flying height (*h*) increases, the anomalies become lower and broader until they merge as a single broad anomaly. Exactly the same effect occurs with increasing depth of burial of anomalous bodies in uniformly magnetic rock for the same altitude of measurement. There is an analogous effect in gravity surveys.

indicating where buried magnetic bodies are located, they can also show how they have been displaced by faults. Buried faults show up as straight features in the magnetic anomaly pattern of an area. Because faults are often related to mineralization (Part I, Section 5.3) airborne magnetic surveys are very useful reconnaissance tools.

The two geophysical exploration methods we have described so far, those seeking gravity and magnetic anomalies, do not give any precise information about the composition of rocks. Moreover, although they can detect anomalous areas, indicate the shape of anomalous masses of rock and delineate geological structures, anomalies revealed by them give little indication of size or depth of possible ore deposits. Consequently they are useful in the reconnaissance stages when regional gravity or aeromagnetic maps are used, but not in detailed exploration, unless no other ground-based methods give meaningful results. More often than not, detailed geophysical surveys exploit the electrical properties of rock, where results *can* be interpreted in terms of depths and sizes of the bodies responsible for anomalies.

4.3   Electrical surveys

The ability of a material to conduct electrical currents is expressed by the ratio between an electrical potential difference (measured in *volts*[A]) and the resulting current (measured in *amperes*[A]) that is induced to flow in the material. This ratio is the material's electrical *resistance*[B] (measured in *ohms*[B]). This is usually expressed as a material's resistance to a current passing through a $1\,m^2$ surface area for a distance of 1 m, and is termed its *resistivity*[B] (measured in *ohm metres*[B]). Resistivity is the most variable of physical properties, and for rocks and minerals other than metals has a range from 10^{-5} ohm m (good conductor) to 10^{13} ohm m (very poor conductor), over 18 orders of magnitude (Table 2). The reason for this huge range is basically the fundamental difference between the molecular structures of minerals (Block 1, Section 2.2.2). The values in Table 2 suggest, at first sight, that measurement of rock resistivity should provide good discrimination between barren rocks and those containing high proportions of oxide and sulphide ore minerals. There is, however, a complicating factor.

| resistance | ohms |
| resistivity | ohm metres |

Rocks invariably have interconnected pores or cracks in them. If these contain water in which ions are dissolved, a low resistivity medium, then the rock resistivity will be decreased below that of the rock-forming minerals. Porous or fractured rocks in wet climates can therefore have resistivities as low as 10 ohm m, values comparable with those rocks containing small but possibly economic amounts of oxide or sulphide ore minerals. Greater concentrations of these ore minerals will produce much lower resistivities, however. So direct measurement of resistivity is only successful for finding high-grade ore deposits. It is also used in surveys for groundwater (Block 4, *Water Resources*).

4.3.1   Resistivity surveys

Resistivity surveys are made by applying a voltage to two *electrodes*[A] buried in the ground. This causes a current to flow between them, penetrating into the underlying rocks. Figure 24 shows how electric currents flow in uniformly resistive rock. The lines of current flow are at right angles to lines of equal *electrical potential*[A]. If the electrodes are long parallel wires laid horizontally, instead of vertical rods, then a rectangular pattern in plan view will be produced between them. This is the principle of the now outmoded *line electrode method*[B]. The potential at the surface between the electrodes is measured at regular intervals using two search electrodes and a voltmeter to produce a map showing lines of equal potential. Current can flow more easily in an orebody of low resistivity and so the potential difference across it would be less than across resistant barren rock. The lines of equal potential would then be more widely spaced over the orebody. For a body of relatively poor conducting properties the reverse would apply: the lines of equal potential would bunch closer together while current flow-lines diverged around it to remain at right-angles to lines of equal potential.

line electrode method

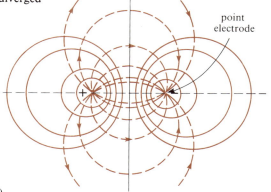

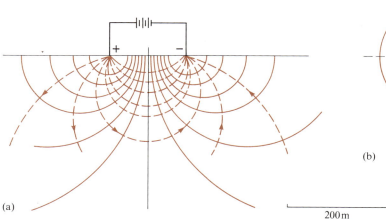

(b)

200 m

(a)

Figure 24 (a) Cross-section and (b) plan views of lines of current flow (dashed) and equal potential (solid) developed in rocks of uniform resistivity from point electrodes.

ITQ 14 If Figure 25 had been the result of a line electrode survey over uniform rock it would have shown a rectangular grid. Use the irregular shape of the lines of current and equal potential on Figure 25 to outline areas underlain by rock of low and high resistivity.

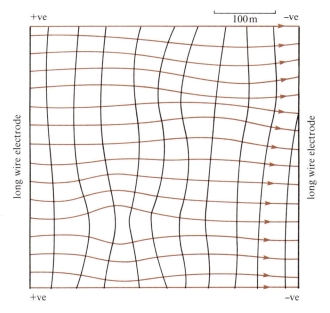

Figure 25 Plan view of lines of equal potential (black) from a line electrode survey over an area of varying resistivity. Lines of current flow (brown) at right angles to equal potential lines are shown to aid interpretation.

There are two basic types of modern resistivity survey, one which examines the variation of resistivity with depth, or *vertical sounding*[B], and another which examines lateral variations, or *horizontal profiling*[B]. In both cases an array of four

vertical sounding

horizontal profiling

31

equally spaced electrodes is used (Figure 26a), two to apply an electric curent and two across which the electrical potential is measured. The spacing of the current electrodes partly determines the depths to which current will penetrate and flow. For very closely spaced electrodes, the current does not penetrate very deeply. The wider the spacing, the deeper some current can flow. As a result more deeply buried rocks can have an effect on the resistivity of the system and so be detected if they are anomalous.

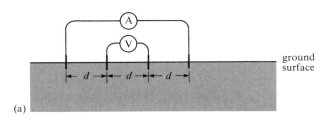

(a)

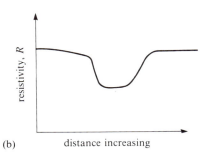

(b)   distance increasing

Figure 26 (a) Arrangement of current (A) and potential (V) electrodes used in horizontal profiling of resistivity. (b) Plot of resistivity with position on a survey line across a resistivity anomaly.

In horizontal profiling, the depth of penetration is kept constant by keeping the same spacing (d in Figure 26a) between the current electrodes and the potential electrodes. The electrodes are moved as an array along the line of survey. Both the current and the potential are measured at each station so that the resistivity can be calculated and a graph of resistivity against distance can be plotted, as in Figure 26b. A change in measured resistivity identifies qualitatively a boundary at a fixed depth between two different rock types, one of which has a lower resistivity than the other and possibly contains a high proportion of sulphides. Clearly, horizontal profiling is useful in *detecting* the presence of orebodies, and would be suitable early in an exploration programme as it is simple to perform.

Vertical sounding is used to measure the depth to a body detected by horizontal profiling. The arrangement of current and potential electrodes is the same as shown in Figure 26a, but the centre of the array is kept fixed instead of being moved along a survey line. The separation of the four electrodes (d) is increased in steps to increase the depth of current penetration, so that changes in resistivity with increasing depth are monitored. Results are plotted as resistivity against electrode separation (Figure 27). A measure of the depth to the low resistivity anomaly is made by comparing this curve with standard curves produced by mathematical models.

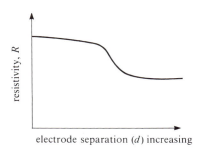

electrode separation (d) increasing

Figure 27 Plot of resistivity against electrode separation from a vertical sounding survey above a buried sulphide body detected by horizontal profiling.

4.3.2 Spontaneous polarization surveys

Under certain climatic conditions sulphide ores near the surface are oxidized during the early stages of secondary enrichment (Part I, Section 4.3); however, the same ores create strongly reducing conditions at depth because of their content of sulphides and the absence of oxygen. Oxidation produces free electrons, by reactions such as $Fe^{2+} = Fe^{3+} + e^-$, which flow as small currents in the low resistivity sulphides. These currents are balanced by a flow of ions in pore fluids (groundwater) in the surrounding rock to complete the circuit. This is analogous to a galvanic cell or simple battery and is the basis of *spontaneous polarization (SP) surveys*[B].

spontaneous polarization (SP) surveys

On Figure 28 the lines of spontaneous current flow are widely spaced in high resistivity country rock, and 'bunch up' at the electrically polarized end of the sulphide body. The lowest electrical potentials are associated with the most closely

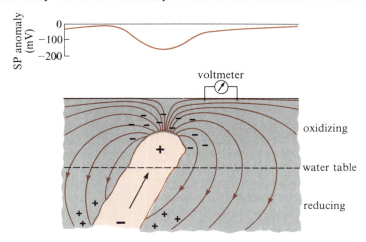

Figure 28 Spontaneous polarization of a sulphide orebody being oxidized above the water table. Free electrons cluster around the upper surface of the sulphide body, and a current flows within the sulphides and in the surrounding, more resistive country rock.

32

spaced lines of current flow and vice versa. For a sharply bounded sulphide body, therefore, the potentials change abruptly just above the body so that lines of equal potential too are tightly bunched. SP surveys measure the surface electrical potential in an area using a pair of electrodes and a voltmeter, usually on a grid basis (e.g. Figure 29).

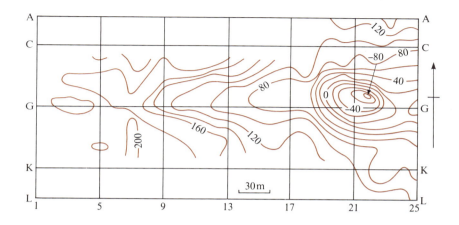

Figure 29 A map of lines of equal potential resulting from spontaneous polarization. The letters and numbers around the margin (and the lines) are from a survey grid. The contours are in millivolts.

ITQ 15   Where on Figure 29 are natural oxidation reactions most likely to be affecting buried sulphide orebodies?

4.3.3   Induced polarization surveys

Electrolysis[A], where oxidation and reduction are *induced* by an applied potential, is the basis of *induced polarization (IP) surveys*[B]. An electrical current is passed through rock, and at the surface of sulphide orebodies some of the sulphides become oxidized as in spontaneous polarization. When the current is switched off, this oxidation reaction continues for a short while so that the orebody continues to be electrically polarized and a potential can still be recorded by search electrodes. In a general sense, then, IP survey data can be interpreted in the same way as those from SP surveys.

induced polarization (IP) surveys

4.3.4   Electromagnetic surveys

If long wavelength radiation passes through a conducting body, alternating electrical currents are induced within the conductor. This, in principle, is how an aerial converts broadcast radio waves into electrical currents which a radio receiver uses to produce sound. The converse is illustrated by radio transmission, when alternating currents flowing in a conductor cause long wavelength radiation to be emitted. Since sulphide and oxide mineral deposits are good conductors of electricity it is possible to exploit these phenomena and detect them by *electromagnetic (EM) surveys*[B].

electromagnetic (EM) surveys

In one variant, an area is 'illuminated' by long wavelength radiation produced by a transmitting coil on an aircraft. Conducting bodies beneath the surface have alternating currents induced in them — they act as aerials. These currents, in turn, produce secondary radiation which may be detected using a receiving coil towed behind the aircraft or held by another field worker. In recent years EM surveys have been developed to use the very long wavelength radiation employed for secret communication with nuclear submarines and produced naturally by lightning strikes. In both cases the radiation passes directly through the Earth and is therefore more convenient than the two-coil method, because only a receiving coil is required. The details of EM surveys are extremely complex so we discuss them no further. However, you should note that they form an important means of reconnaissance exploration. Like gravity and magnetic data, those from EM surveys are difficult to interpret in terms of both size and depth of anomalies, and they are not used frequently in detailed surveys. Costs and efficiencies are roughly the same as those quoted in Section 5 for aeromagnetic surveys (Table 3).

The geophysical and geochemical data collected by the methods described in Sections 3 and 4 relate to many abstract properties. Because geologists rely on only one sense, that of sight, to relate these data to geological maps, it is equally important to present these data in a visual form. This forms the subject of the second part of AV6.

4.4   Summary of Section 4

1   Geophysical surveys seek buried bodies of rock with anomalous physical properties which are expressed in anomalous measurements of various force fields at or above the surface. They may rely on natural force fields as in gravity, magnetic and SP surveys, or they may apply energy to the Earth to induce force fields, as with resistivity, IP and EM surveys.

2   Gravity, magnetic and EM methods are used primarily in reconnaissance surveys. Resistivity, SP and IP methods are those most commonly used in detailed surveys.

3   Detailed interpretation of geophysical anomalies relies on models that predict variations in their shape resulting from various orientations, depths and shapes of anomalous bodies. In reconnaissance, anomalies are usually interpreted in map format and are used to detect subsurface geological features such as faults and to distinguish buried masses of rocks of grossly different chemical composition.

Objectives and SAQs for Section 4

Now that you have completed Section 4 you should be able to:

1   Define in your own words or recognize valid definitions of the terms and concepts introduced or developed in this Section and listed in Table B.

9   Describe the principles behind the interpretation of anomalies defined by gravity, magnetic and electrical surveys. *(ITQ 13)*

10   Select the type of geophysical survey most useful for detecting different kinds of mineral deposit.

11   Using data from a geophysical survey, identify anomalies and make qualitative estimates of the shape and depth of the bodies responsible for them. *(ITQs 14 and 15)*

Now do the following SAQs.

SAQ 10 *(Objectives 1 and 10)*   Which of the geophysical survey techniques (i)–(iii) should be capable of detecting (a) massive copper–nickel sulphide bodies in gabbro, (b) tin-bearing veins related to granite buried beneath limestones, (c) magmatic segregation deposits of chromite in a gabbro? (Use Table 2.)

(i) Gravity survey.   (ii) Magnetic survey.   (iii) Electrical survey.

SAQ 11 *(Objective 1)*   Explain in one sentence how SP surveys differ from IP surveys.

SAQ 12 *(Objectives 1 and 11)*   Figure 30 shows a total field magnetic anomaly in Cumbria where the magnetic inclination is at about 45° to the south. Is the anomaly due to (i) a buried magnetic sphere, (ii) a horizontal magnetic sheet, (iii) a fault, or (iv) a combination of (i) and (iii)?

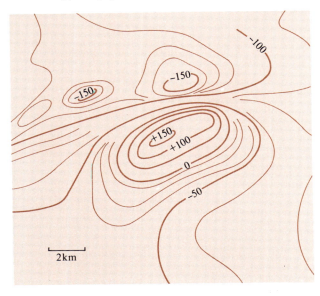

Figure 30   Anomaly map of the total magnetic field of an area in Cumbria. The contours are in nanoteslas.

SAQ 13 *(Objectives 1 and 9)* A long vertical pipe of rock with a high content of magnetic minerals will have one magnetic pole close to the surface, the other very deeply buried. Sketch the shape of the magnetic anomaly if the Earth's field inclination is vertical, using the method in ITQ 13.

5 Evaluation

Study comment In this Section we consider the information that is needed before the high capital costs of opening a mine can be committed to an ore deposit found by exploration. You should remember that the gathering of geological information continues through this phase and into actual operations.

The results of an exploration programme identify geologically, geochemically and geophysically anomalous areas, but they do not necessarily confirm the presence of either reserves or resources. Much more specific data are needed before the economics of mining can be assessed. This means estimating the grade and mass of the body, which give an idea of the value of contained metal when related to its market price. Other measurements relating to the details of shape, depth and geological setting have to be made in order to plan mining (see Section 6). The nature of the ore, what minerals it contains, and their grain size and properties, must be examined, as these will determine the ore processing techniques (see Sections 7 and 8). Finally, all the factors relating to the geographical location of the deposit must be examined in detail to evaluate the logistics of opening a mine.

5.1 Geological considerations

Various means can be used to measure an ore deposit, including trial pits and trenches. However the most commonly used is drilling, where samples of subsurface rock are obtained for analysis. This can simply use rotary drilling where chips of the rock being cut are blown to the surface with an air blast; this costs about 20 US dollars per metre (1982). A complete record is acquired using a hollow diamond-tipped bit which cuts a continuous cylinder of rock, known as a *drill core*[B], that is periodically recovered and stored for analysis; this costs about 40 US dollars per metre (1982).

drill core

ITQ 16 A porphyry copper deposit outcrops over an area of 1 km^2 and its secondarily enriched zone extends to a depth of 200 metres. Estimate the cost of diamond drill coring at a spacing of 100 metres to evaluate the enriched part of the deposit. How does this compare with the cost of reconnaissance exploration (satellite remote sensing plus airborne EM surveys and interpretation) over the surrounding 900 km^2 and with ground surveys (detailed geological mapping and soil surveys) over the surrounding 100 km^2? Table 3 contains a summary of the costs and efficiencies of different exploration techniques. Assume that each area is a square and that the flight lines of the EM survey were parallel to both sides and spaced at 1 km intervals.

Although your answer to ITQ 16 was based only on estimates, it emphasized the critical nature of the decision to proceed from exploration. In practice, a limited amount of drilling and trenching may accompany the later stages of exploration in order to add subsurface information about geology and geochemistry to the results of geological, geochemical and geophysical surveys (Figure 2). It is an inevitable expense in estimating the exploration risk more precisely.

As drilling is so expensive, the location and pattern of drill holes need to be planned carefully to ensure they have a good chance of passing through a suspected ore deposit. This planning is based partly on detailed geological mapping, and partly on the shapes of anomalies. Figure 31a shows an example of drilling to prove most efficiently the existence of a vertical hydrothermal vein underneath a symmetrical and linear lead anomaly. A vertical hole on the anomaly would stand a good chance of completely missing, therefore inclined holes are

Table 3 Approximate costs and average efficiencies of some exploration methods (1982 data)

Methods	Cost/US dollars	Efficiency/km² day^{-1}
Preliminaries (1:100 000 scale)		
satellite remote sensing	0.02 km^{-2}	$> 10^6$
interpretation and map	0.7 km^{-2}	10^4
aerial photograph remote sensing	10 km^{-2}	500
interpretation and map	5 km^{-2}	50
airborne geophysics (magnetic and EM)	25 km^{-1}	500
interpretation and map	10 km^{-2}	25
literature search	150 day^{-1}	—
Field studies		
geological reconnaissance	200 km^{-2}	10
detailed geological mapping	600 km^{-2}	1
geochemical orientation	10 km^{-2}	50
drainage survey	50 km^{-2}	25
soil or biogeochemical survey	750 km^{-2}	2
resistivity and SP	50 km^{-1}	10
IP	5 000 km^{-1}	0.5
diamond drill cores	40 m^{-1}	—
shaft sinking	5 000 m^{-1}	—

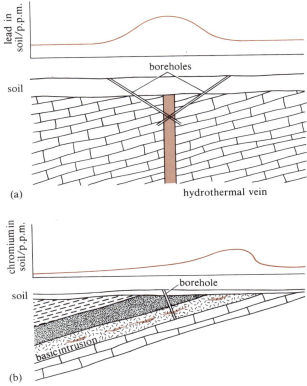

(a)

(b)

Figure 31 The location and orientation of boreholes to detect most efficiently (a) a vertical hydrothermal vein suspected from a symmetrical lead anomaly in soil; (b) a sheet of magmatically segregated chromite in a gently dipping stratiform gabbroic intrusion suspected from an asymmetrical chromium anomaly in soil.

drilled towards the anomaly from one side. For most other types of deposit vertical holes may suffice. However, in the case of a dipping stratiform body (Figure 31b), which will not exist beneath part of an area due to erosion, the location of the drill holes must be decided on the basis of knowledge of the direction of dip.

Once a deposit is known to exist from initial drilling a carefully designed pattern of new holes is used to construct a three-dimensional map of the deposit. Figure 32 shows how the drilling of a major copper deposit may have proceeded as more and more data became available. The holes are numbered in order and data are shown for the grade of the ore and the depths of the top and bottom of the deposit. You should note how some of the later holes are closely spaced to estimate the trend of the high grade parts of the deposit, while some are 'stepped out' from known areas to seek new extensions of the ore on the basis of discovered trends.

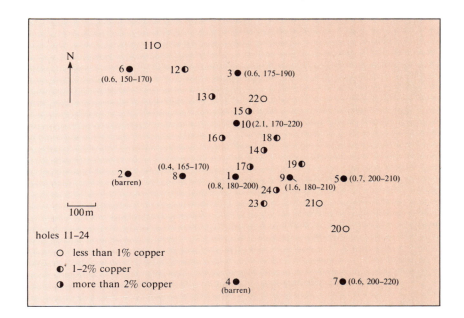

Figure 32 A map showing the location of boreholes, in order of drilling, during the early stage of evaluating an area showing anomalous copper in soil. Data on grade (per cent), and depth (metres) to the top and bottom of a mineral deposit are shown for the first ten boreholes (solid black dots), and symbols relating to grade for the rest.

As well as supplying the drill core, from which the shape, volume, and grade of a deposit can be estimated from chemical analyses and measurements of ore thicknesses (known as *blocking out*[B]) core samples provide the raw data for other studies important in designing a mine. They may be used to determine the strength of the ore and surrounding rock, on which depends the safe construction of the mine. These are the geotechnical investigations on Figure 2. Information on how the rock breaks is needed to plan the blasting and primary ore crushing. The most important information required for planning ore processing concerns the properties of the individual ore minerals, their grain size and how they relate to one another and to waste material in the ore, all of which can be evaluated at an early stage in planning by examining core. Even such mundane factors as the bulk hardness of the ore, on which depends the wear and tear on mine equipment and how much this will cost the operation, can be determined during evaluation. Together with definition of the shape of the orebody, drilling can reveal geological hazards, such as faults and the depth to the water table.

As well as being affected by these various geological parameters, a working mine faces a whole variety of risks (Block 1, Section 5.2.3). A mine designed for profitable operation only at full capacity is vulnerable to unforeseen risks. One of the primary functions of evaluation is therefore to provide detailed information that will enable mining to be as flexible as possible — it must be profitable during lean periods as well as during booms. Ideally, the evaluation identifies blocks of high grade ore to be worked when times are bad, blocks of 'run of the mill' ore for normal operations and blocks of varying grade and properties that can be blended to ensure that the processing plant continues operation even when the main blocks are closed down for some reason. However, the opening of a mine does not depend just on the existence of an ore deposit, a suitably high metal price, and favourable geological conditions. Its planning also depends on *where* it is located and the logistics of operation.

5.2 Logistic considerations

It is impossible to provide an exhaustive list of non-geological factors which have some influence over the economics of a mineral deposit, but the following questions give an indication of the kinds of information needed.

How accessible is the area? Will a new road, railway or port have to be built? Is there an electricity supply, or will a power station have to be built? Will the mine have to be self-sufficient, or can food and spares for equipment be bought locally? Can operations continue throughout the year or does the climate prohibit work during some seasons? What laws govern pollution control during operations and eventual restoration of the mine area? Are there sites where waste can be dumped cheaply, safely and legally? Who are the owners of the adjoining land and will they hinder development? Is there an indigenous skilled workforce, or will a mine township have to be built for immigrant labour? What wages need to be paid to

blocking out

attract suitable personnel to the mine? The list can go on and on, even to investigating the best lubricant for surface vehicles under the particular climatic conditions.

Only when all the economic, geological and other considerations have been evaluated can the final decision to mine be taken. The financial commitment after this decision is enormous compared with all the previous work. Because site preparation is essentially an engineering exercise we skip it and continue in Sections 6–8 with a discussion of the geological factors involved in winning the target metal.

5.3 Summary of Section 5

1 Evaluation of a discovery consists of acquiring geological data relating to the mass, grade, shape and engineering properties of the mineral deposit, and non-geological information about the geographic, economic, political and environmental considerations that relate to the feasibility of mining it profitably.

2 The geological data are obtained by drilling, and analysing rock cores, which is more expensive than previous exploration. On this information are based designs for the mine itself, the processing plant, waste disposal and the ways in which different grades of ore can be mined to give the operation maximum flexibility.

Objectives and SAQs for Section 5

Now that you have completed Section 5 you should be able to:

1 Define in your own words or recognize valid definitions of the terms and concepts introduced or developed in this Section and listed in Table B.

12 Estimate the costs of evaluation programmes and compare them with those relating to exploration and mine development. *(ITQ 16)*

13 Use drill core data to assess the depth, shape and value of mineral deposits.

Now do the following SAQ.

SAQ 14 *(Objective 1 and 13)* Figure 33 shows an array of nine boreholes (perspective effects have been removed and vertical and horizontal scales are equal). Table 4 gives depths, thicknesses and grades of the core samples where they intersect with a mineral deposit. Plot these data on Figure 33.

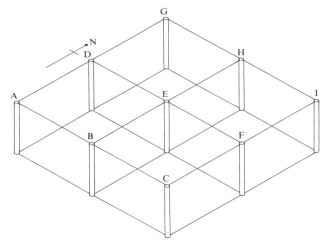

Figure 33 A three-dimensional projection of nine vertical boreholes used in blocking-out a mineral deposit. The holes are 500 metres apart and 300 metres deep.

Table 4 Borehole data for Figure 33.

Hole	Depth of ore /metres	Grade (per cent copper)
A	50–100	0.3
B	100–250	0.8
C	100–200	3.0
D	50–100	0.1
E	50–150	0.4
F	50–70	0.7
G	0–30	0.01
H	200–300	0.2
I	250–300	0.2

(a) What shape is the orebody and are there any obvious geological problems?

(b) Using the average grade and thickness, estimate the value of the block between holes B, C, E and F (average density of ore = 3 000 kg m^{-3}, price of copper = £800 per tonne).

(c) Where would you concentrate further evaluation drilling?

MINING, MINERAL PROCESSING AND SMELTING

6 Mining methods

Study comment We come now to the technological aspects of ore deposits, which simply mean getting the potentially useful and profitable parts of a deposit out of the ground and into the industrial system as cheaply and efficiently as possible. In this Section we look at how both the nature of ore deposits and the economics of mining determine the method of extraction.

Obviously the easiest way to mine is to operate at the surface, merely picking up the ores that are required where they are found in outcrops. The earliest mining was as simple as that, and it was soon followed by the digging of small surface pits. The early development and rise to dominance of true underground mining was inevitable for a host of reasons. First, only rich ores could be used in primitive smelting techniques. Such ores, the *confined deposits*A, are rare and usually occur in narrow bodies, such as hydrothermal veins, which have small surface outcrops and hence limited surface pickings. The lack of any earth-moving equipment other than picks, shovels and animals, meant that large volumes could not be excavated. Underground mines following the intricacies of vein deposits were in operation during the period of the Roman Empire. However, the depth of underground mining was limited by the level of the water table, below which water invaded the excavations. Until simple pumps had been devised, mines could penetrate no deeper. A further depth limit, that imposed by ventilation problems, had to await the invention of means to circulate air. The ultimate limit in modern mines is that imposed by explosive failure of rock under extremely high pressures, and high temperatures caused by the *geothermal gradient*A.

The Industrial Revolution, with its rapidly increasing demands for a growing mass and variety of metals, triggered great improvements in mine design. The invention of powerful explosives and mechanized earth-moving machines enabled large surface excavations to be opened quickly and cheaply. This transformed mineral deposits with much lower proportions of ore (*dispersed deposits*A) than those mined underground from mere anomalies into profitable and huge reserves. In the twentieth century surface mining has rapidly overtaken underground mining as the major source of metals, but subsurface operations still play a substantial role because confined, high grade deposits give a fast return on investment.

Section 6.1 considers how a decision is made to mine a deposit at the surface, below ground or not at all. This decision is based on economics. The project that began as an abstract exploration objective and proceeded through several phases of exploration, with decreasing risk and increasing cost (Figure 2), has still not earned any money. By the time a mine is operating, 8 to 10 years may have passed and more than 10 million US dollars may have been spent. The mine design must therefore ensure rapid returns on this investment, continuous operation and flexibility to take account of changing metal prices.

6.1 Choosing a mining method

The first consideration in deciding how an ore deposit will be worked is to compare the value of the deposit with the cost of extracting the ore. The total cost of transforming one tonne of ore in the ground into a saleable commodity is termed the *unit cost*B. In a grossly simplified form the unit cost (C) is made up of: **unit cost**

C_O = the day-to-day cost of excavating and transporting one tonne of ore

C_w = the day-to-day cost of excavating and disposing of the mass of waste (barren rock or unprofitable ore) associated with one tonne of ore

C_f = the overheads per tonne of ore extracted per year — *fixed costs*B **fixed costs**

C_p = the cost of processing one tonne of ore to a saleable product.

Therefore

$$C = C_O + C_w + C_f + C_p \qquad (1)$$

The first two terms are reasonably simple to evaluate, being made up of costs of explosives and fuel, maintenance costs of equipment, costs of replacement equipment, and wages. The overheads, though, are a nearly endless list, including the cost of long-lived equipment, permanent constructions, land, insurance, environmental restoration and road or rail access. The overheads are divided by the expected lifetime of the mine and the anticipated annual output of the mine in tonnes of ore to give C_f. The cost of processing, C_p, involves separation of ore minerals from waste, smelting and transport of refined metal.

Can you see how the contribution of fixed costs to the unit cost can vary drastically?

If for some reason, perhaps strikes or a decrease in demand, mine output decreases, then the fixed costs increase, because they are the overheads divided by annual output. On the other hand, if productivity makes a surge forward the contribution of fixed costs to the unit cost diminishes.

There are two main economic differences between surface and underground mining. The first relates to the fixed costs, which are higher in underground mining partly because the permanent constructions and long-lived equipment include fixed systems of access shafts and underground roadways and provision for ventilation and drainage. Moreover, usually smaller masses of ore are mined underground than at the surface. In a surface mine, the access is part and parcel of ore excavation itself, and drainage, though necessary, requires only pumps and temporary pipes. As you will see in Section 6.3, underground mines, with all their requirements for safety and precision, require far more sophisticated design and expensive equipment, with consequent additions to fixed costs. The second difference relates to C_O and C_W (equation 1). Both are much greater in underground mines than in surface mines. So much greater, that the ratio of C_f to $C_O + C_W$, which is a measure of the long-term capital investment compared with day-to-day running costs, is in fact lower in underground mines than surface mines.

6.1.1 Mining cut-off

The other side of the economic equation to unit cost is the value of one tonne of ore, the *unit value*[B]. It is much simpler to estimate, the variables being the metal content of the ore — its *grade*[A] — and the price of the metal. Thus:

unit value

$$\text{unit value} = \text{grade} \times \text{metal price} \qquad (2)$$

For any mine, whether surface or underground, there is a 'break-even' point where unit cost equals unit value. Any lower unit value would result in an economic loss, so the *mining cut-off*[B] is defined by this 'break-even' point. For a stable metal price, this can be expressed as the lowest grade of ore that can be mined at a profit, the *cut-off grade*[A]. Because cut-off grade varies directly with the unit cost, and inversely with metal price, it is more useful to consider the *cut-off value*[B], which is obtained by multiplying the grade by the metal price.

mining cut-off

cut-off value

For the purpose of economic analysis, it is possible to divide ore deposits into three categories.

1   Those in which the metal contents merge almost imperceptibly into those of ordinary rock. These are dispersed deposits, examples of which are porphyry copper deposits and some *chemical precipitates*[A] in sediments.

2   Those in which there is a sudden decrease in the metal content of rock at the boundary of the deposit. These are confined deposits, examples being *pegmatites*[A], *magmatic segregations*[A], and *hydrothermal veins*[A].

3   Those which contain parts with low grade mineralization plus patches of high grade ore, which can be termed *locally enriched deposits*[B]. Examples are *residual deposits*[A] and *secondarily enriched deposits*[A].

locally enriched deposits

If we were to plot the variation of grade in an ideal dispersed deposit against the tonnages of ore with different grades, the graph would be like that in Figure 34. The very highest grade ore (with the highest unit value) would have a very small tonnage, and the grade decreases down to the metal content of common rock with virtually infinite tonnage. The curve is of the exponential type. This is similar to Figure 5 in Part I, but in this case the horizontal axis refers to reserves of ore, not of metal. This is a more convenient description of the deposit for mining purposes. For graphs of this kind the cut-off grade depends on the previously fixed cut-off value, and they can be used to estimate the tonnage of ore reserves in a deposit. On this tonnage depend the lifetime of a mine, its optimum rate of production, annual turnover and profits. Grades A and B on Figure 34 have associated reserve tonnages of X and Y. If grade A is the cut-off grade for a particular metal price, and that price increases, a new, lower cut-off grade, such as B, will come into play.

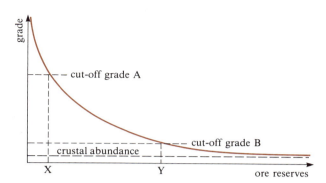

Figure 34 Idealized relationship between grade and ore reserves tonnage in a dispersed ore deposit.

This lower grade ore now has the estimated cut-off value for the deposit. In the case of dispersed deposits a price rise will dramatically increase the reserves. Conversely, the reserves will fall alarmingly if the price falls. Should the price remain constant, but the unit costs increase, perhaps due to inflation or unforeseen mining difficulties, then the cut-off value has to increase to balance unit costs, the cut-off grade increases and the reserves decrease dramatically. The opposite trend would prevail for a decrease in unit costs, due perhaps to increased mining efficiency.

You should remember that dispersed deposits are never uniform, and changes in the cut-off grade inevitably mean changes in the shape of the orebody as well as changes in reserves. This also means changes in the mine design.

In reality prices and unit costs fluctuate together, making the cut-off grade rather ephemeral. Demand may also vary. To cope with these complexities, companies mining dispersed deposits generally extract ore with all grades at a constant rate, and carefully stockpile uncrushed ore when demand is low. Stockpiled ore can then be blended at any time to meet both increasing demand and the changing cut-off grade before being processed. Another strategy is to mine selectively, which is easier at the surface than underground. It is therefore essential that, for a mine in a dispersed deposit, the company should always retain sizeable reserves of high grade ore for this purpose, rather than working it out for short term, high returns.

Figure 35 shows the grade–tonnage relationship for a hypothetical confined deposit (Part I, Figure 70). The ultimate tonnage is clearly defined, unlike that of a dispersed deposit

> **ITQ 17** Look at Figure 35. (a) Are the reserves in a confined deposit more or less sensitive to fluctuations in price or unit cost than those in a dispersed deposit?
> (b) Estimate by what factor the metal price has to decrease, or the unit cost to increase, for the reserves to fall to half the tonnage indicated by the cut-off grade in Figure 35. (Remember that cut-off grade varies directly with unit cost and inversely with price.)

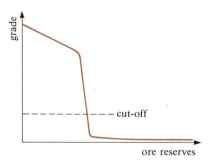

Figure 35 Idealized relationship between grade and tonnage of ore reserves in a confined ore deposit.

The answer to ITQ 17 implies that high-grade confined deposits can be mined continuously and profitably at constant rates even when prices and costs fluctuate dramatically. These are sometimes called 'bonanza' deposits. However, they come in all shapes and sizes, as shown in Figure 36, and can occur at any depth. Consequently the unit costs for each are different, so the cut-off grade for a particular metal varies from deposit to deposit. The three in Figure 36, which have the same gross value, are arranged in increasing order of unit cost from (a), which can be mined by surface working, to (b), which must be mined underground by simple methods, to (c), which can only be mined by complex underground methods.

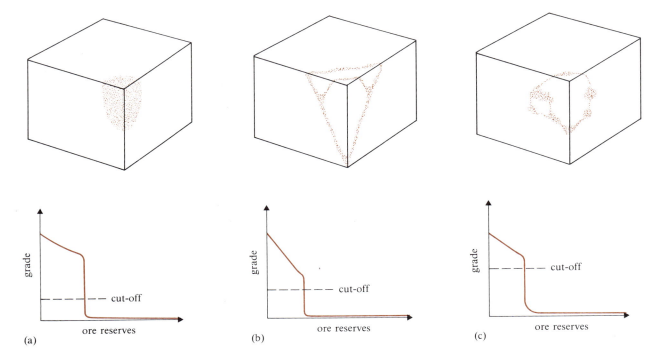

Figure 36 Block diagrams to illustrate idealized relationships between grade, ore reserves tonnage and cut-off grade, and geological sketches for different confined ore deposits. (a) Massive deposit; (b) simple vein deposit; (c) irregular vein deposit with larger ore pockets. The density of brown dots reflects the ore grade.

ITQ 18 Figure 37 shows a schematic diagram of a dispersed deposit after secondary enrichment. Sketch a grade–tonnage curve for this deposit and comment on the implications for the reserves of decreasing the unit costs if the metal price remains constant.

Thus, planning a mine depends on considerations of economics and technology, on the geometry and location of the deposit relative to the surface, and on the way grades vary within the deposit. The simplest plans to understand are those related to surface mining.

6.1.2 Stripping ratio

Because the unit cost of surface mining is generally lower than that of underground operations, surface working is favoured where possible. Often the choice is obvious, as Figure 38 shows. Figure 38a shows a horizontal ore deposit covered by a thin *overburden*A of valueless rock. This deposit can be mined from the surface in just the same way as limestone for cement or clay for bricks (Block 2), depending on the *overburden ratio*A. However, Figure 38b shows an ore deposit that reaches the surface — an obvious candidate for surface mining — but extends to very deep levels. This implies that a decision has to be made about how deep surface mining can extend before it is more economic to open an underground mine in the deeper parts of the deposit. Figure 38c shows a thin horizontal deposit beneath thick overburden. The decision here is between surface mining only or underground mining only, depending on the unit value of the ore.

Surface working of deposits that are not horizontal entails the removal not only of the overburden but also of the waste around the ore deposit, as you will see.

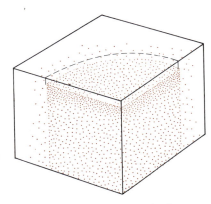

Figure 37 A geological sketch of a dispersed ore deposit with a secondarily enriched zone. The density of brown dots reflects the ore grade.

Figure 38 Different types of ore deposit. (a) Shallow stratiform; (b) inverted cone; (c) deep stratiform.

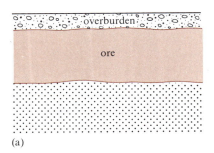

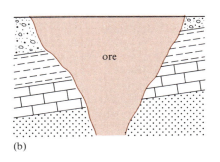

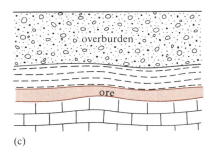

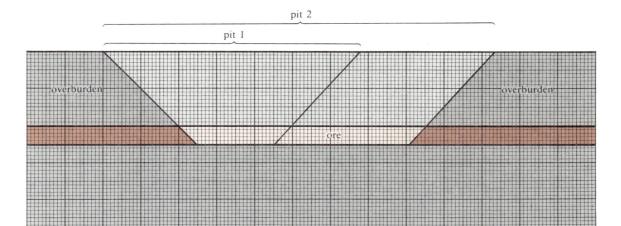

(a)

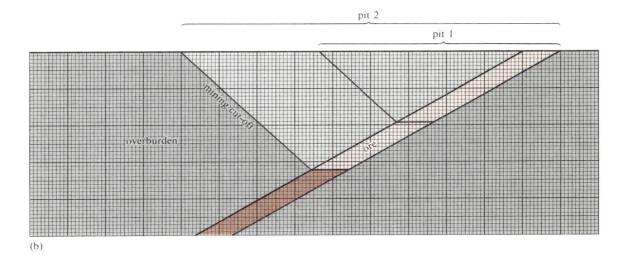

(b)

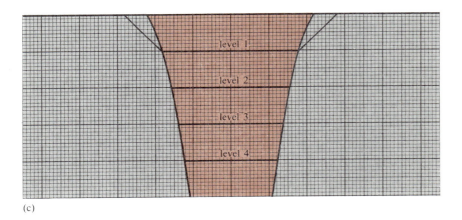

(c)

Figure 39 Open pits in (a) a shallow horizontal layer; (b) a dipping layer; (c) a conical ore deposit. Note that areas on these cross-sections are proportional to volumes (and hence masses) of rock.

To quantify the economics of surface mining requires a term more general than overburden ratio. This is the ratio of the mass of waste that must be removed during mining to the mass of ore that is extracted, and is known as the *stripping ratio*[B]. This tells us about the relationship of C_W to C_O (equation 1). Most surface mines have the form of an inverted, truncated cone known as an *open pit*[B]. The reason for this shape is that even very strong rocks will collapse if vertical faces are cut in them. To ensure stability of the open-pit walls they are never steeper than 45° from horizontal. To explain how this affects the stripping ratio, here are a few examples.

stripping ratio

open pit

In the simplest case, mining a horizontal layer (Figure 39a), the excavation of a pit with 45° walls involves only the removal of overburden and ore. The areas on Figure 39 are proportional to volumes and, if the density is constant throughout, to masses as well. Therefore the stripping ratio is very close to the ratio of overburden thickness to thickness of ore. As the pit grows in size the stripping ratio *decreases* slightly, as you could confirm by comparing the areas of overburden and ore for pit 1 and pit 2.

Figure 39b shows a dipping layer of ore with two pits.

> Does the stripping ratio stay constant as the size of the pit increases?

No, it increases as deeper parts of the ore are mined.

Figure 39c shows a cross-section of a conical ore deposit with a density of $4\,500\,\mathrm{kg\,m^{-3}}$ in barren rock with a density of $2\,600\,\mathrm{kg\,m^{-3}}$. In it are shown four different levels, each of which represents the depth to which ore can be broken by drilling and blasting from the layer above. The stripping ratio for a pit to level 1 is calculated in the following way:

$$\text{volume of ore (proportional to area)} = 386 \text{ squares}$$

$$\text{volume of waste} = 57 \text{ squares}$$

$$\text{stripping ratio} = \frac{\text{mass of waste}}{\text{mass of ore}}$$
$$= \frac{57 \times 2\,600}{386 \times 4\,500} = 0.08$$

> **ITQ 19** Draw on Figure 39c the outlines of pits with 45° walls that expose levels 2, 3 and 4. Using the method of calculation above, estimate the stripping ratio for each stage in the excavation of the open pit.

Clearly the stripping ratio increases with depth for many deposits, but the way it increases depends on the shape of the deposit. A deposit that narrows downwards has a stripping ratio that increases much more rapidly with depth than does a deposit that becomes wider with depth. At a certain depth the stripping ratio becomes so high that equation 1 is dominated by C_W, and the unit cost exceeds the unit value. The mine must either be abandoned or continued by underground methods.

Although the unit cost of underground mining is often higher than that of surface mining, it is not so much affected by the depth of operations. As underground mines go deeper only the fixed costs, for deeper shafts and extra ventilation, increases markedly. A very much lower amount of waste is produced underground, and as we shall show in Section 6.3, most of this is left in the mine and only the ore is brought to the surface.

Figure 40 shows in a general way how the unit costs of surface and underground mining vary with depth. At depth *d* the curves intersect, and deeper still, underground mining is cheaper than surface mining. For ores with a high unit value, surface mining can be replaced by underground operations below *d*. For ores with a low unit value this will not be possible and a *maximum economic stripping ratio*B will be reached at much shallower levels.

Further complications develop where an ore deposit has an irregular shape, and particularly when it contains ore with various grades.

> **ITQ 20** Figure 41 shows a deposit where ore in block A, with an average grade of 3.0 per cent metal, has been mined to the maximum economic stripping ratio. Calculate the minimum grade of ore in block B that is required for surface mining to continue to the base of the deposit. (Assume equal density for ore and waste.)

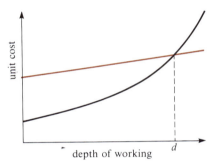

Figure 40 Idealized variation of unit costs of surface mining (black) and underground mining (brown) with depth of working. Depth *d* represents the depth at which surface mining is replaced by underground operations.

maximum economic stripping ratio

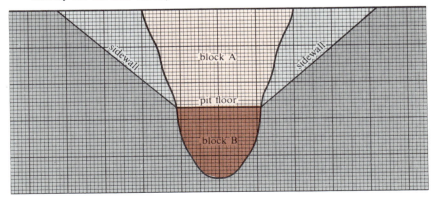

Figure 41 A sketch of an open pit in a pipe-like ore deposit.

6.2 Surface mine design

There are few restraints on the space available in surface workings, other than the size of the deposit. Surface mining therefore benefits from economies of scale. Large capacity earth-moving equipment can be used, with correspondingly higher productivity. This cuts down the unit costs, and low grade deposits can be worked at high tonnages. Some large, low grade porphyry copper deposits (Part I, Section 2.4) are more profitable than small confined deposits. This is because mining of the dispersed deposit can be entirely within rock above cut-off grade whereas the design of pit walls forces waste to be mined with ore in the confined deposit.

Surface mining can be divided into three categories:

1 Open-pit or bench mining of deposits that are deep but of restricted width, such as porphyry copper deposits;

2 Open-cast or strip mining of relatively thin but extensive flat deposits, such as chromite layers in a gabbroic intrusion or chemically precipitated ores in sedimentary rocks;

3 *Alluvial mining*B of unconsolidated near-surface deposits on land or beneath water, such as placer deposits.

alluvial mining

In open-pit mining the waste is stored in spoil heaps, and must not obscure unmined ore, but in the other categories it can be dumped where ore has been removed, and landscaped if necessary.

6.2.1 Open-pit or bench mining

Open-pit mining is usually used for steeply dipping beds or veins, stockworks and pipes, and massive irregular deposits (Part I, Figure 11). Development of the mine is usually quite simple (Figure 42). The overburden is removed from the whole area of the final pit. The pit is then developed sequentially downwards in a series of steps or *benches*B, with walls left at a safe angle and of a height determined by the

open-pit mining

benches

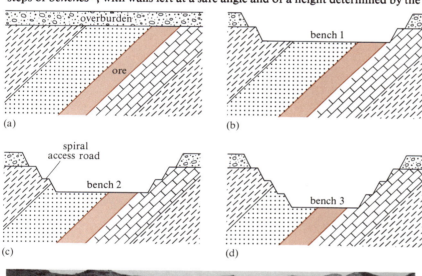

(a)   (b)   (c)   (d)   (e)

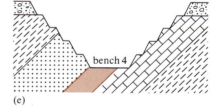

Figure 42   A typical sequence of operations during development of an open pit, showing: (a) stripping of soil; (b) removal of overburden and mining of ore on first bench; (c) development of second bench and access road; (d) and (e) development of deeper benches. You can see that more waste has to be excavated than in our theoretical studies of Figures 39 and 41 in order to prepare the site and develop access roads.

Figure 43   A typical open pit showing benches and access roads.

equipment to be used, usually in the range 10–20 metres. Access to the pit bottom is maintained by haulage roads that spiral down the side of the pit, cutting across the benches as shown in Figure 43. Roads are seldom steeper than about 1 in 10 because there are gradient limitations on haulage equipment. As the pit becomes deeper it is often invaded by water, especially once it has penetrated the water table, so pumps must be installed to keep it dry.

There are three well-defined stages in the excavation of an open pit:

1 *Drilling and blasting*   In many mines work is arranged so that one 8-hour shift is responsible for drilling a series of shot-holes, charging these with explosive and blasting, while the following two shifts remove the blasted material, so that a continuous cycle can be maintained.

2 *Stripping*   Stripping describes the process of collecting the fragments of ore and waste broken by blasting, and removing them from the pit. The most commonly used machines for this are very large electric shovels or front-end loaders. The costs of stripping and loading are surprisingly high, forming about 15 per cent of the total operating costs.

3 *Haulage*   Haulage of ore from the blasting site to the processing plant is the most expensive single item in any open-pit operation, absorbing over 30 per cent of the total costs. The vehicles used are almost exclusively diesel-electric trucks, which have superseded the railways used previously because they are much more manoeuvrable and are cheaper to run.

From the 1960s onwards there has been a dramatic increase in the average size and total capacity of trucks used. The largest now carry over 300 tonnes of ore and are so big that their drivers are often perched 5 metres or more above ground level. The bigger the truck, the fewer journeys it has to make, and the fewer the trucks, the fewer drivers and maintenance personnel needed, and the lower the unit costs. Such large trucks are naturally very expensive, and their maintenance is a big problem. One of the mining engineer's most important responsibilities is to keep the trucks running.

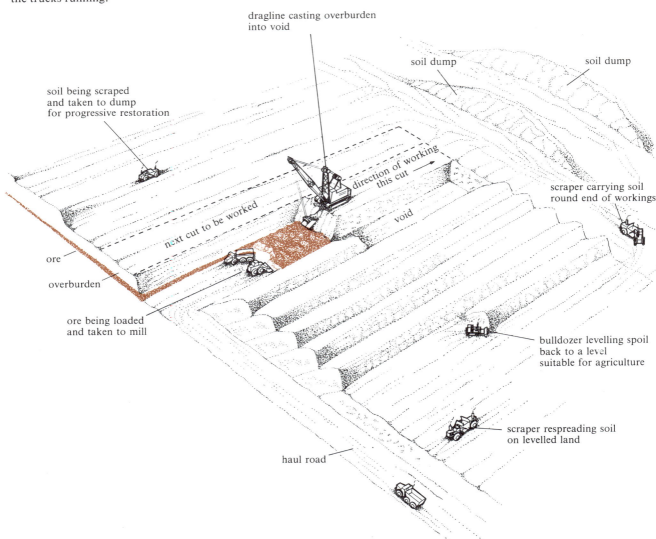

6.2.2 Open-cast mining

In *open-cast mining*[B] (sometimes known as strip mining) waste is dumped inside the worked-out section of the excavation. The method is almost invariably confined to horizontal or gently dipping stratified deposits near the surface. Mining proceeds laterally instead of downwards. Initial development consists of excavating a trench or cut through the overburden to expose the first strip of ore. Only the overburden from the first cut is dumped outside the mining area. Once ore has been extracted from the first cut, overburden from the second strip can be discarded into it, and so on.

The most important factors in design are the stripping ratio, the cut-off value in a dipping deposit, and the distance between the point where overburden is loaded and the point where it is to be dumped. Mining is simplest where this distance is small and overburden can be cast directly on to the spoil heap by an excavating machine, such as the dragline shown in Figure 44. Drilling and blasting of overburden and ore may be necessary, but much open-cast mining is in material that has been loosened by weathering. The limit on this type of mining is the depth of the ore. At depths below about 50 metres direct casting of waste becomes impossible and some form of benching must be employed.

In open-cast mining, no pit or spoil heap is left when the ore is exhausted, and it is quite simple and cheap to restore the land to its previous use. In open-pit mining, by contrast, the volume of waste is large and, to prevent the ore from being obscured, it is dumped some distance from the pit. Restoration is therefore almost as large and long an operation as mining itself, and the costs are very high. The major problem is that the broken rock occupies a greater volume than it did before mining because it cannot be packed tightly unless in a finely ground form.

6.2.3 Alluvial and marine mining

Various types of mineral deposit, such as placer deposits, *black sands*[A], *manganese nodules*[A] and metal-rich marine muds comprise unconsolidated sediment at the land surface or on the sea floor. Mining them does not involve drilling, stripping or blasting, merely scraping them up. Problems arise in (i) scraping up sufficient tonnages cheaply to ensure constant flow through the processing plant, (ii) operating in water-saturated or submarine deposits.

In a dry deposit conventional earth-moving equipment is often sufficient, but with waterlogged surface deposits working may be by bucket line in a flooded pit (as seen in many sand and gravel pits). In more extensive and rich deposits such as the tin-bearing gravels of Malaysia and gold-bearing stream placers, floating dredgers (Figure 45) are employed; these often carry ore processing equipment and immediately return waste to worked-out 'lagoons'.

Submarine mining, particularly in the deep ocean, is not yet established commercially, but various techniques are being considered such as continuous bucket loops or pumps mounted on ships, which will also carry equipment to concentrate the ore.

Figure 45 A floating dredger recovering gold from stream placers near Fairbanks, Alaska.

6.3 Underground mining

Underground mining is much less flexible than surface working. The time taken to reach the orebody is usually greater, and it is much more difficult to deviate from the planned production rate. The maximum output from any working place is usually much lower than in surface mining, and it is often necessary to mine simultaneously in several locations to achieve the desired production target. It is therefore extremely important that the overall mine plan and the work schedules are carefully prepared well in advance of actual operations. Consequently the lead time is greater than for surface mining. During production, mining must be co-ordinated with the development of horizontal, vertical or inclined passages to give access to particular blocks of ore and to sustain constant production.

The design of an underground mine depends on three factors: how best to reach the ore; how to ensure that a minimum of waste is mined; and how to ensure safe operations. The last two depend on the shape, attitude and mechanical properties of the ore deposit.

6.3.1 Access

Access from the surface to an ore deposit is by two routes: (i) vertical *shafts*[B]; (ii) *declines*[B] (slopes, *drifts*[B]) — gently sloping tunnels usually inclined between 5° and 20°. These allow people, materials, air and power to reach the working places and people, ore, waste rock, water and used air to be brought to the surface. The choice of access depends on the geometry and position of the deposit. Shafts are normally used to gain access to depths exceeding 300 metres. Declines allow heavy equipment to be driven directly in and out of the mine, but require more excavation than shafts, because of limits on their gradient.

shaft
decline   **drift**

A minimum of two entries into the mine is necessary for safety. These form a ventilation circuit, fresh air entering at one opening and contaminated air leaving from the other. They also provide an alternative exit in the event of an underground accident. The entry must be close to the orebody to reduce long and expensive underground roadways, but if it is too close a portion of the orebody may have to be left intact to prevent damage to the access.

The engineering requirements can be summed up in two main questions: can the rock and ore support the mining excavations, and can gravity be exploited in designing the operation? The answers depend on four main geological factors:

1 The dip of the deposit, which governs the dip of excavation and the extent to which gravity can be exploited;

2 The shape of the deposit, which limits the size of excavations and determines the kind of support — a very wide deposit will need more roof support than a narrow one;

3 The strength of the ore, which determines whether or not unworked portions can be left safely unsupported, and whether ore can be used to support the roof or walls;

4 The strength of the *host rocks*[A], which determines the stability of the roof and walls of excavations and the need for support.

You saw in Part I that ore deposits have a great variety of shapes and dips. Moreover, ores and rocks vary considerably in their strength. So unlike surface mines, each of which is usually one of three fundamental designs (Section 6.2), every underground mine is unique. There is, however, a means of classifying them based on geology:

Class A Deposits that are wider than they are thick, such as stratiform deposits with a dip of less than 10°;

Class B Deposits with large vertical dimensions, such as large conical or roughly spherical bodies, steeply dipping stratiform deposits, or hydrothermal veins.

The constant aim is for the highest possible proportion of excavations to be in the ore itself. Consequently, in the case of class A, the excavation has a semi-permanent roof and floor made of waste rock, the walls being working faces in ore. In class B it has semi-permanent walls, and the ore in the roof and floor is excavated. Where the walls in class B are not vertical, there is a simple terminology — the wall above the excavation is called the *hanging wall*[B] (it may literally be left hanging), that beneath is the *footwall*[B]. Where the host rock is strong, the roof or hanging wall needs little support; when it is weak, support is essential. In an

hanging wall
footwall

excavation that needs support, if the ore is strong then parts of it can be left temporarily in place to prevent collapse; if the ore is weak then artificial support is needed. We shall illustrate some of the principles and terminology of underground mining by examining a few of the possible types of mine in deposits of classes A and B.

6.3.2 Mines in shallow-dipping ore deposits

Such ore deposits include stratiform chemical precipitates and magmatic segregation deposits, and also salt (Block 2) and coal (Block 5). Mines in them take the form of flat or shallow-dipping (less than 10°) parallel-sided cavities. As the cavities become larger, so the tendency for the roof to sag and collapse increases, depending on the strength of the rock that forms the roof. With a strong host rock, a system of semipermanent cavities can be developed. These give great flexibility for selective working of different parts of the ore deposit depending on changes in the cut-off grade. Some support is needed as mining progresses, and the cheapest means is to leave pillars of unworked ore. The size and spacing of pillars depend on

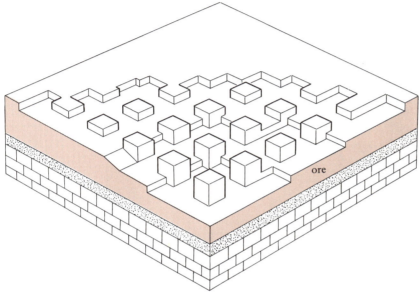

Figure 46 A cutaway diagram of a room and pillar operation with multiple benches in a thick stratiform ore deposit.

the load on the roof and thus on the depth of working, and on the strength of the ore. Only medium to strong ores can be mined in this way, which is known as *room and pillar mining*[B]. Figure 46 shows such a mine in a thick ore deposit, where a series of benches have been excavated downwards from the top of the deposit. Worked-out areas of the mine can be filled by waste rock or by injection of the fine-grained material left after ore processing, thereby cutting the costs of shifting waste rock to the surface and its subsequent disposal. Moreover, the waste-filled cavities become stabilized, allowing the progressive removal of ore originally left in the pillars.

room and pillar mining

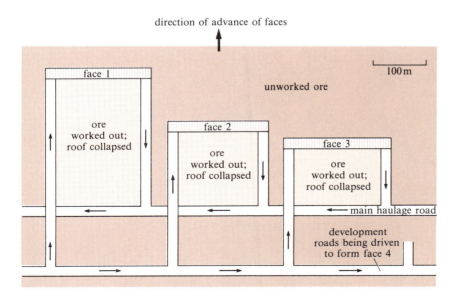

Figure 47 A plan view of a typical longwall mine. Arrows show the flow of ventilating air.

Where both host rock and ore are mechanically weak, collapse of the roof can be prevented only by expensive artificial supports. In such cases the deposit is worked by a series of faces or longwalls up to 200 metres long, which are established some distance from the shaft and are driven towards it. The roof close to the face is artificially supported by props, which are removed as the face moves, allowing the roof to collapse in the worked-out area (Figure 47). Access, ventilation and ore-removal roadways are maintained by artificial support. In such *longwall mining*[B], the ore, being weak, is generally produced in small fragments which can be cheaply transported to the surface by conveyor belts.

longwall mining

6.3.3 Mines in deposits with large vertical extents

Where an ore deposit extends over a great vertical distance in a large massive body or a steeply dipping sheet, gravity can be used to aid transport of ore within the mine and in some cases to break the ore under its own weight. Figure 48 illustrates some aspects of mine design in a vertical tabular deposit. The ore deposit is divided horizontally and vertically into several blocks (Figure 48a). Each of the vertical divisions corresponds to a mining level. Access to each level is provided by a main haulage roadway or drift running horizontally to one side of the deposit and parallel to it. This drift is joined by a large *crosscut*[B] to the main shaft, thus giving a means of moving ore, equipment and air to and from the deposit. Small crosscuts to the deposit from the haulage drift correspond to the horizontal divisions between blocks of ore, and provide minor access from which the sequential mining of individual blocks proceeds and along which ore is transported (Figure 48b). In the case of a dipping tabular deposit the permanent haulage roads are located beneath the deposit so they do not collapse into the worked deposit.

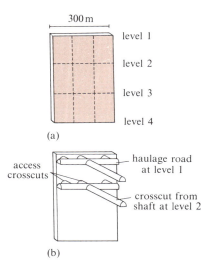

(a)

(b)

Figure 48 The basic operations in the working of a vertical tabular ore deposit. (a) The division of ore into blocks; (b) crosscuts driven towards the orebody from the main shaft, main level haulage drifts driven parallel to ore deposit, crosscuts driven from each main level drift to ore.

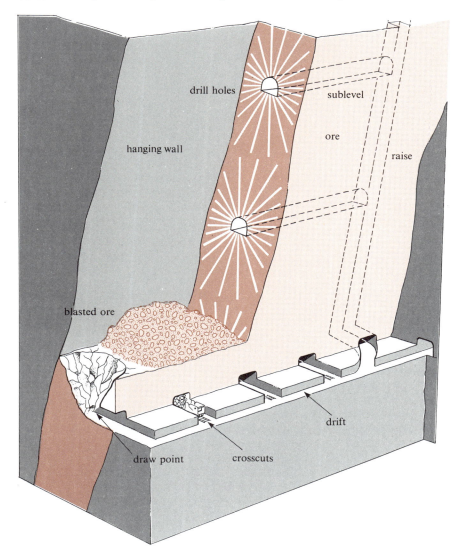

Figure 49 A cutaway of a partially worked sub-level stoping operation. Access to the sublevels is provided by a raise which has been bored upwards from an access crosscut, through the ore. The brown tones indicate the ore, the grey is barren rock.

Figure 49 shows in more detail how the mining of a block proceeds. Access within the block is achieved by vertical *raises*[B] and horizontal sub-levels. The sequential mining of blocks opens up cavities called stopes and is therefore known as *stoping*[B]. It can proceed in three main ways depending on rock strength.

raise

stoping

Where medium to strong ores are surrounded by strong host rock, the worked-out parts can be left as self-supporting cavities. Ore is first extracted from the base of a block to give a working floor, into which are driven conical *draw points*[B]. These funnel the broken ore into crosscuts and the transportation system. The stope is then expanded upwards by blasting the roof and higher parts of the block from sub-levels. The ore falls to the stope floor and is mucked-out into the drawpoints. This is called *sub-level stoping*[B], which is the simplest type.

draw point

sub-level stoping

Where the surrounding rock is weak, it is impossible to maintain an open stope because the wall would collapse. As stoping proceeds upwards, the cavity must be filled with waste rock from higher levels or fine-grained waste from the processing plant. The working floor then rises as the stope is extended upwards. Channels, or *ore passes*[B], down the filled stope must be kept open by iron or concrete collars to allow ore to pass down to the main haulage drift (Figure 50). This method is known as *cut-and-fill stoping*[B]. Because a safe room must always be maintained in the stope, this method can only be used with strong ores.

ore passes

cut-and-fill stoping

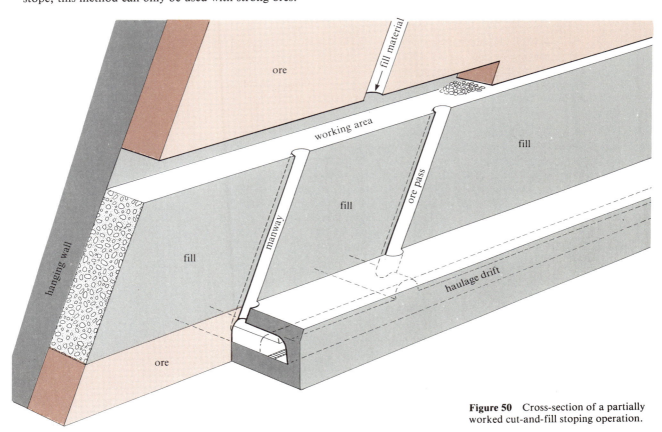

Figure 50 Cross-section of a partially worked cut-and-fill stoping operation.

Where both the host rock and the ore are weak, a working floor with drawpoints is developed as in sub-level stoping but with artificial support. The block of ore is then blasted and allowed to collapse under its own weight to the base of the stope, where it is drawn off. Because it always occupies a greater volume than solid ore, the broken ore supports the walls (Figure 51). This method is called *shrinkage stoping*[B], and it has three main disadvantages. First, a high proportion of ore is 'tied-up' in the stope and remains so for the lifetime of the mine. Second, wall rocks mix with ore and dilute it. Third, broken ore can jam in the stope and cannot be reached safely.

shrinkage stoping

In ore deposits with large dimensions horizontally as well as vertically, even more complex designs are necessary as the transportation network must be within the ore itself. Often very large caves are opened, thereby undercutting blocks of ore. Drawpoints are built in the cave floor and they lead to a transportation system in a level beneath the floor. These methods are known as *caving*[B], and in some cases the overlying land surface subsides in spectacular fashion. Caving methods are most appropriate for large deposits where either the ore or the host rock is weak.

caving

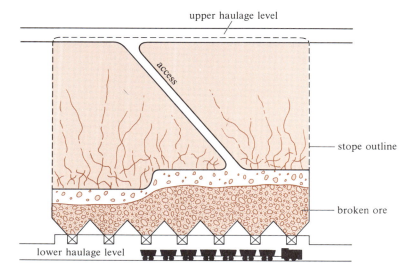

Figure 51 Cross-section of a shrinkage stope.

6.4 Summary of Section 6

1 The choice of a mining method depends first and foremost on the economic factors, setting the unit value of the deposit against the unit costs of extracting the ore. Unit costs relate to transport of ore and waste and also to the overall capital tied up in permanent mine constructions set against the annual extraction rate and the lifetime of the mine. All the economic factors are variable to a greater or lesser degree.

2 The decision on whether to operate by surface mining depends on how deep excavations can be before the costs of removing waste outweigh the value of the ore. This is expressed by the stripping ratio. Below the maximum depth of surface workings, ore must either be abandoned or mined by underground methods. Because less waste has to be moved in underground mining and gravity can be used as a 'free' energy source in some cases, the predominant contribution to underground costs is the high fixed capital investment involved in building the mine. Consequently unit costs are not increased as much by depth in underground mining as in surface mining and underground operations are more economic for some deposits of large volume and/or high grade.

3 Designing a surface mine is simple, the only important geological consideration being the strength of the rock, which determines the slope angles of the pit walls. Most effort centres on tailoring the mine to the capabilities of mechanized equipment.

4 Underground mine design is strongly conditioned by the strength of the ore and the host rock, by the shape and thickness of the deposit and by variations in grade. Sub-level stoping is appropriate for strong ores in strong host rocks; cut-and-fill is for strong ores in weak host rocks; shrinkage stoping is for weak ores.

Objectives and SAQs for Section 6

Now that you have completed Section 6 you should be able to:

1 Define in your own words or recognize valid definitions of the terms and concepts introduced or developed in this Section and listed in Table B.

14 Recognize and evaluate the various economic, geological, technical and other factors controlling the viability of surface and underground mines. *(ITQs 17–20)*

15 Identify the geological conditions suitable for surface and underground mining and their relationship to operating methods.

Now do the following SAQs.

SAQ 15 *(Objectives 1 and 14)* Which of the following types of deposit has reserves that are most sensitive to variations in the cut-off grade? (i) Nickel sulphide in a magmatic segregation deposit; (ii) low grade copper–lead–zinc precipitates in sediments; (iii) a residual deposit of nickel in the capping to weathered peridotite.

SAQ 16 *(Objectives 1 and 14)*   Figure 52 shows a cross-section through a pipe-like ore deposit. The maximum economic stripping ratio has been calculated as 2.0.

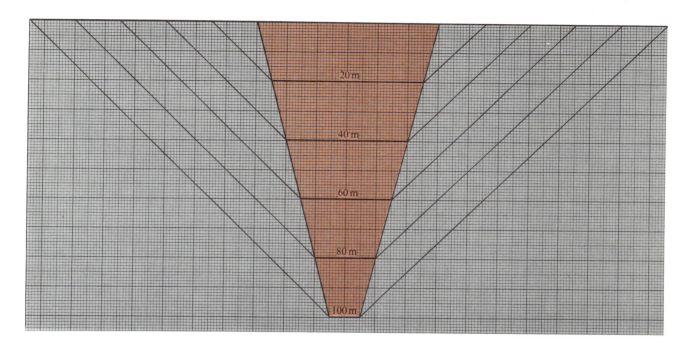

(a) Determine the maximum depth of an economic open-pit mine, on the following assumptions: (i) the maximum safe slope of an open pit is 45°, (ii) the ore and the waste have the same density, (iii) the pit is developed in 20 metre benches, and (iv) the ore has a constant grade.

(b) How much higher must the grade be in the deepest portion of the pit where the stripping ratio is greater than 2.0 if the deposit is to be worked down to 100 metres?

Figure 52   Cross-section of an ore deposit.

SAQ 17 *(Objectives 1 and 15)*   Identify the methods of underground mining appropriate for the following ore deposits.

(a) A strong iron-ore bed, 2 metres thick, dipping at 5° and overlain by medium-strength shales.

(b) A steeply dipping, strong quartz vein containing gold which, varies from 3 to 20 metres in thickness in a low-strength host rock.

(c) A massive porphyry copper deposit with high strength surrounded by weak rocks.

(d) A weak nickel sulphide layer dipping steeply within strong gabbroic rocks.

7   Mineral processing

Study comment   The theme of this Section is the need to concentrate ores as cheaply and efficiently as possible. We survey the preliminary information required to choose an optimum sequence of ore processing techniques, and look at economic and environmental factors that restrict the siting of ore processing plants. We then study the principles and practice of rock-breaking, an energy-intensive process which requires maximum efficiency to avoid excessive cost. We also look at sizing methods, which ideally should produce an appropriate size-range of particles for separation procedures and ensure that no energy is wasted on over-reduction of particle size. We see how differences in physical and chemical properties between minerals form the basis of several techniques for mineral separation, and consider the practical and economic limitations governing the choice of separation methods. Most procedures result in a wet mineral concentrate, so lastly we examine methods of removing water cheaply.

7.1 Basic operations

Ores extracted by mining still have the same grade as when they were in the ground. Low grade ores, such as those of dispersed copper deposits, may contain over 99 per cent of *gangue*[A], and even high grade ores often contain a lot of gangue. The *place value*[A] of raw metal ores can be higher than that of the cheapest aggregates, and it is rarely economical to transport such ore very far (cf. Block 2, Section 1.4.1). Mined ore must therefore be processed to remove as much gangue as possible leaving a concentrate of the ore mineral. *Concentrates*[B] are a saleable commodity, relatively easily transported, and in an acceptable form for *smelting*[B], the process by which usable metals are obtained from ore concentrates. No two ore deposits contain precisely the same mineral assemblage, and the proportions, textures and grain sizes of minerals vary, even across the same orebody. This means that no set of operations is optimal for more than one combination of minerals.

concentrates
smelting

Figure 53 shows three general stages of mineral treatment. Few processes work perfectly, so unsuitable material is recycled to earlier stages. The three main steps are:

1   *Liberation*[B], which involves freeing ore minerals from gangue, usually by breaking the rock into fragments smaller than the mineral grain size. The aim, which is seldom completely achieved, is that each fragment should consist of only one mineral. Many ores are composed of interlocking crystals, and liberation of individual mineral particles is difficult, because fracture is more likely to occur across crystals than between them (see Figure 27a, Block 1).

liberation

2   *Separation*[B], which is the process of sorting liberated particles into groups, each containing a single mineral, by exploiting their physical or chemical differences.

separation

3   Drying procedures, which eliminate water from the processed ore.

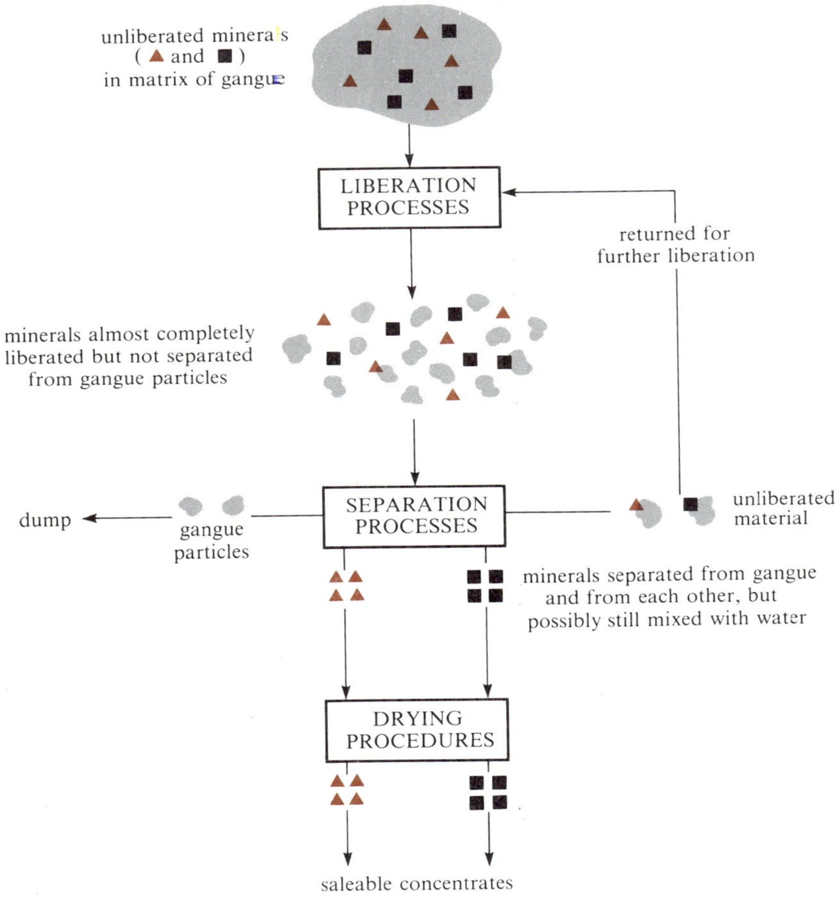

Figure 53   Schematic representation of mineral processing.

flowsheets

Mineral processing engineers use *flowsheets*[B] to represent what goes on inside a processing plant, thus avoiding laborious written descriptions. Simple flowsheets (Figure 53) show only the movements of solid materials. More advanced flowsheets also show fluid flows, flow-rates, addition of reagents, exact type and size of equipment, and even such details as conveyor belts.

7.1.1 Mineral identification and distribution

Before deciding on an appropriate sequence of processing techniques, a mineral processing engineer needs to know precisely which minerals are present, and the proportion, grain size, and distribution of each. A variety of microscopic, chemical, X-ray, and other techniques are available for mineral identification, which needs to be thorough, for the presence of trace elements in a mineral may significantly alter its properties. Mineral grain size dictates the particle size to which ore must be broken to achieve liberation. If grain size varies across the orebody, modification of processing techniques may be necessary as extraction proceeds. In some ore deposits, the metal sought may be distributed amongst several minerals, not confined to one alone.

> **ITQ 21** Ore from a tin mine contains, by weight, 98 per cent gangue, 1.5 per cent cassiterite (SnO_2), and 0.5 per cent stannite ($Cu_2S.FeS.SnS_2$). Stannite contains 28 per cent of tin.
>
> (a) Calculate the percentage of tin in cassiterite (the relative atomic mass of oxygen is 16 and that of tin is 119). (You may wish to refer to ITQ 3 in Part I.)
>
> (b) Calculate the total tin content of the ore, as a percentage.
>
> (c) If a mineral treatment plant were set up to recover only the cassiterite, what would be the maximum amount of tin recovered (as a percentage of the tin content of the original ore)?

When an element is distributed between different minerals there are inevitable losses. Despite the high price of tin, an additional separation process to recover tin from subsidiary minerals is rarely economically worthwhile.

7.1.2 Siting of processing plant

Low grade ore has a very high place value, and cannot be economically transported very far. Therefore, ore concentration has to take place very close to the mine. If water, power and transport are not available, it is necessary to provide them, although obviously it is advantageous to site the processing plant as near as possible to these services. Equally obviously, the plant should not be built over ore reserves.

Higher grade ores produced from small mines present special problems. Consider an area containing many small hydrothermal veins, each rich in galena (PbS). Is it worth building a processing plant on each site? Alternatively, is the place value of the ore low enough for the output from several mines to be transported to a central processing plant, with consequent economies of scale? These and other questions need careful consideration before a processing plant is built.

By contrast, ore concentrates have a low place value and can be transported appreciable distances, and hence the location of a mine imposes no constraint upon the siting of a smelter.

7.2 Principles of mineral liberation

Many metallic ores occur as hard compact rocks, and are not easily broken into individual grains. On the other hand, *unconsolidated*[A] placer deposits contain the minerals as separate grains, so no liberation is necessary. Tin mining in Nigeria and Malaysia operates on very low-grade ores because no energy or expense is required to liberate the minerals. Most minerals are liberated by crude mechanical methods which break the rock along random planes. This method seems straightforward, but in the example in Figure 54 only two ore mineral particles are liberated by four random breakage lines. In general, no significant liberation occurs until the particle size of the broken rock is smaller than the size of the original mineral grains. Complete liberation can be achieved only if the rock is broken into very small fragments. However, this results in the breakage of already liberated grains, thereby wasting energy, increasing processing costs, and producing too many *fines*[A]. Particles of the required size are removed from the system as soon as possible, usually by *screens*[A]. A balance must be struck between achieving maximum liberation, minimum fines, and an optimum grain size for subsequent separation procedures.

The energy used in breaking rocks represents almost half the cost of a mineral product, hence efficient rock-breaking methods are sought. Some of the

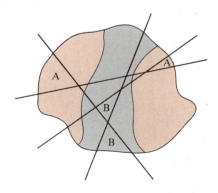

Figure 54 The effect of random breakage lines on a particle. Only two ore particles (A) and two gangue particles (B) are obtained. The other fragments all contain both ore and gangue. (Remember that particles are three-dimensional.)

properties of a hard rock can be investigated by a *'stress[A]/strain[A]'* test (Figure 55). OA on the figure shows the stress range over which the rock suffers no irreversible structural changes. Repeated stresses within this range will not break the rock, and all the applied energy is wasted. AF is the range of stresses within which irreversible structural damage occurs. Repetition of such damage eventually fractures the rock, but wastes energy. Practical rock-breaking machines are designed to apply a direct stress slightly greater than that corresponding to the failure point F, so that the rock will break immediately.

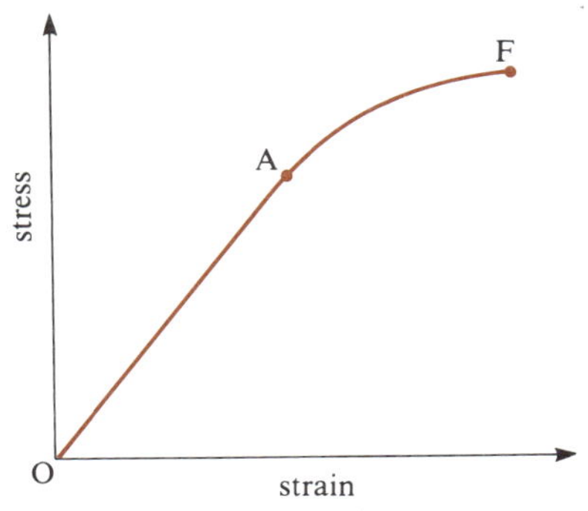

Rocks are always found to be weaker than expected from theoretical models, because regions of weakness are introduced by 'foreign' atoms and molecules, and by cracks and pores between grains. The energy required to achieve a certain size reduction can be related to the size of the initial feed particles by the empirical equation:

$$W = 10 W_i \left[\frac{1}{\sqrt{P_{80}}} - \frac{1}{\sqrt{F_{80}}} \right] \tag{3}$$

where W = the electrical input required in kilowatt hours per short ton (a short ton = 970 kg)

P_{80} = the size in micrometres (μm) below which 80 per cent of the product lies

F_{80} = the size in micrometres below which 80 per cent of the initial feed lies

W_i = the *Bond work index*[B], which is the work in kilowatt hours per short ton required to reduce the particle size of the rock from infinitely large to a product size of 100 μm diameter.

Figure 55 Stress—strain curve for a rock specimen.

Bond work index

The value of the Bond work index is determined by a laboratory test, and varies from about 8 for soft but coherent rocks to about 15—18 for hard compact rocks.

ITQ 22   The Bond work index of a rock is 12 kWh per short ton. Calculate the energy in kilowatt hours needed to achieve a ten-fold reduction in size (a) if the initial feed has an F_{80} value of 100 μm, (b) if the initial feed has an F_{80} value of 1 000 μm. What can you deduce from your results?

7.3   Comminution

Comminution[B] in mineral processing means the breaking of rock fragments. Breaking rocks down to 1 or 2 cm diameter is traditionally called *crushing*[B]. Further breakage to smaller particle sizes is called *grinding*[B].

comminution
crushing
grinding

Primary crushing produces rock fragments of about 5 cm diameter. The type of crusher varies, but a typical example is a *jaw crusher*[B] (Figure 56). Secondary crushing reduces rock fragments to about 1 cm diameter. In any one machine, the *reduction ratio*[B] of feed particle size to product particle size is about 4:1. Several types of secondary crushers are used. Figure 57 illustrates a *cone crusher*[B]. In primary crushing the rock fragments follow a simple linear path, or *open circuit*[B], being sequentially reduced in size by a succession of crushers. Secondary crushing systems often contain *closed circuits*[B] from which material already sufficiently crushed is screened off, and over-sized material is returned for recrushing. Guard magnets are usually included in crushing systems to remove objects such as broken steel tools which otherwise could damage the crusher.

jaw crusher

reduction ratio
cone crusher
open circuit

closed circuit

In grinding, rock fragments of 1—2 cm diameter are broken to a size at which an acceptable degree of liberation is achieved. As the answer to ITQ 22 shows, the

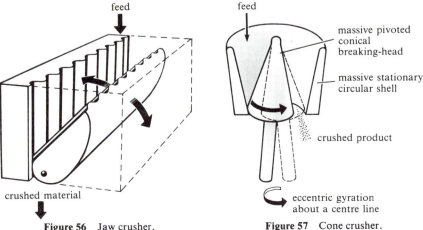

Figure 56   Jaw crusher.   **Figure 57**   Cone crusher.

energy requirements for grinding rise steeply as the particle size becomes smaller. Grinding is usually carried out in rotating metal drums, containing heavy steel rods or balls, which fall on to the rock as the drum rotates. Coarse grinding is usually done in open circuit in *rod mills*[B]. Finer grinding is done in *ball mills*[B] (Figure 58), working in closed circuit. It is advantageous to grind ores in the wet state, because the incompressibility of water aids the extension of cracks so that less energy is used. If a dry product is required, dry grinding is practised, because the thermal energy needed to dry the product would exceed the energy saved by wet grinding.

rod mill   **ball mill**

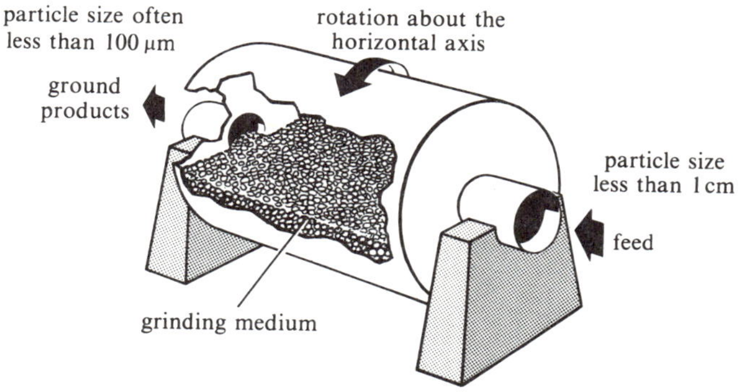

particle size often less than 100 μm

rotation about the horizontal axis

ground products

particle size less than 1 cm

feed

grinding medium

Figure 58   Ball mill.

7.4   Sizing the output from comminution

As it is important to remove particles from a comminution circuit as soon as they are the correct size, each circuit is designed to give an output below a certain size (Figure 59). If mineral particles are passed through a screen having a known aperture size, particles larger than that aperture will be retained on the screen.

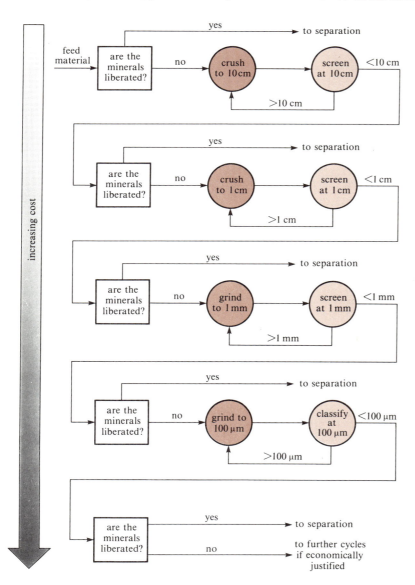

Figure 59   Generalized representation of a liberation system.

Screening is usually restricted to particles above 1 mm diameter. Whether a material is screened wet or dry depends on its characteristics. In dry screening the small particles may adhere to the larger ones, but wet screening helps to disperse fine particles and so produces better results. Various designs of screens are used. Figure 60 illustrates a cylindrical rotating screen.

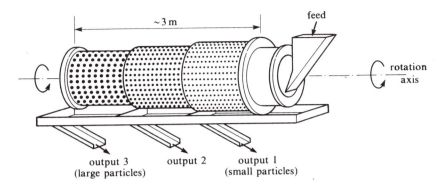

Figure 60 Cylindrical rotating screen.

Classifiers[B] sort particles into different size groups by using the speed at which they settle in a fluid. Larger particles fall through a liquid faster than smaller particles of the same mineral. The velocity (V) reached by a sinking spherical particle is given by the equation:

$$V = \frac{(\rho_p - \rho_l)D_p^2 g}{18\eta} \qquad (4)$$

classifiers

where D_p = the diameter of the particle (m)
  ρ_p = the density of the particle ($kg\,m^{-3}$)
  ρ_l = the density of the liquid ($kg\,m^{-3}$)
  g = the gravitational acceleration ($9.8\,m\,s^{-2}$)
  η = the viscosity of the liquid ($kg\,m^{-1}\,s^{-1}$)

Suppose that two sizes of particles of the same mineral are dropped into a column of water, and that the larger particles reach velocity V_1 and the smaller reach velocity V_2. Now imagine that the water is made to move upwards with velocity V (chosen to be greater than V_2 but less than V_1) by feeding more water into the base of the column. Large particles will sink with velocity $V_1 - V$, whereas the small particles will be carried away up the column (they will rise with velocity $V - V_2$). The column has separated the two sizes, and thus has acted as a classifier. A *cyclone*[B] is a classifier in which the settling forces are amplified by centrifugal motion, thus speeding separation (Figure 61).

cyclone

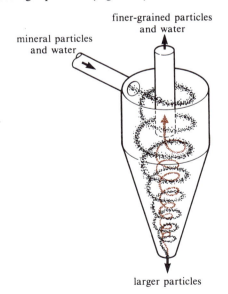

larger particles

Figure 61 Cyclone separator.

7.5 Mineral separation

The concentration of metals is just as low in liberated ore as in untreated rock. Separation of valuable minerals from gangue and from each other is often difficult, and total recovery of the economic mineral is rarely achieved. The separation techniques discussed here are based on simple properties (grain shape,

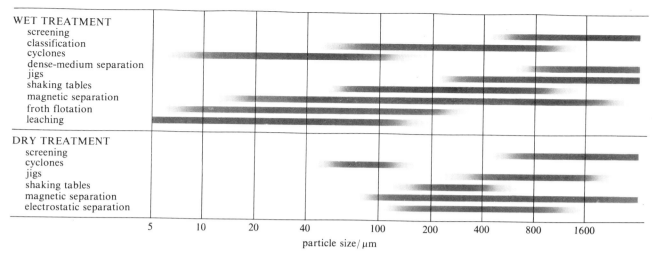

Figure 62 Particle size ranges for various mineral separation processes.

appearance, density); on electromagnetic properties (magnetism, resistivity); or on chemical differences (surface properties, *chemical reactivity*[A]). We examine the techniques in the same order as a processing engineer would consider their use, that of increasing complexity and cost. Figure 62 shows the range of particle sizes appropriate to each of the techniques. You will find it useful to refer to as you work through this section. You will see that some treatments appear under both 'wet' and 'dry' headings. Dry processes usually use less energy, but wet processes are often used because water aids dispersion of small particles and acts as a lubricant for larger particles.

7.5.1 Mechanical sorting

Manual sorting, the earliest method of mineral separation, is still used where labour is cheap, rock fragments are over 4 cm diameter, and differences between mineral and gangue are obvious. Similar tasks are performed by *mechanical sorters*[B]. Electronic sensors examine a mineral fragment from several directions, so most of the surface is 'seen'. A property such as colour or fluorescence is compared with a chosen standard, and particles not meeting the standard are rejected by mechanical plungers or air jets. Fragments from about 20 cm to 4 mm diameter can be sorted by these methods.

mechanical sorters

7.5.2 Float–sink methods

It is easy to separate sawdust from sand by placing a mixture of the two in a bucket of water. In the same way, it is possible to separate pairs of liberated minerals in a fluid of intermediate density, chosen so that one mineral floats and the other sinks. Hence the method is called *float–sink separation*[B]. Because most minerals are heavier than water, heavy organic liquids are used for mineral separations in the laboratory, but are too expensive or too toxic for use on an industrial scale. Suspensions of finely divided (about 50–100 μm) solid particles in water produce fluids having 'apparent' densities much higher than water. Such pseudo high-density liquids are called *dense media*[B] and are prepared from finely ground high-density solids such as ferrosilicon (an alloy of silicon and iron). Ferrosilicon is highly magnetic and thus can easily be recovered for reuse.

float–sink separation

dense media

Density differences as small as 100 kg m^{-3} can be exploited to separate two minerals by the float–sink method, and the techniques are used on particles from about 10 cm to 1 mm diameter (Figure 62). It is imperative that the particles being separated are much larger than the particles used to form the dense medium; if they were similar in size, they would settle out together and a further process would be required to separate them. Larger mineral particles can be separated in open baths of dense media, whilst finer sizes (say 1 cm to 0.6 mm) are separated in dense-medium cyclones, in which the settling forces are increased by centrifugal motion.

ITQ 23 Refer back to equation 4 to answer the following questions.

(a) Why are the ferrosilicon particles not separated along with the mineral fraction in a dense-medium cyclone?

(b) Calculate the *ratio* of the two settling rates, given that the ferrosilicon particles have density 6.5×10^3 kg m^{-3} and diameter 100 μm, the mineral particles have density 3.5×10^3 kg m^{-3} and diameter 1 mm, and the apparent density of the ferrosilicon suspension is 3×10^3 kg m^{-3}.

7.5.3 Shaking methods

'Panning' for gold (Figure 63) is a well known method of mineral separation, and consists of swirling water over a mixture of gold and sand grains. The light sand grains are washed away, and the dense gold is left in the bottom of the pan. Jigs and shaking tables are developments of this principle, and are used to separate minerals with density differences of $1 \times 10^3 \, \text{kg m}^{-3}$ or greater. A *jig*[B] is a water-filled box in which mineral particles are supported on a screen. When water is

jig

Figure 63　'Panning' for gold.

pulsed up and down through the screen and the overlying mineral bed, the particles separate into layers, with denser particles towards the bottom, and lighter particles at the top. Continuous separation of dense from light particles is obtained by removing the heavy minerals through the screen, and the lighter particles from the top (Figure 64). Jigs are usually used for separating particles of about 0.25 to 2 mm diameter (Figure 62), although particles of up to 10 mm can be separated by this method.

Figure 65 shows the behaviour of a particle in a film of water flowing down an incline. The velocity of the water is small in contact with the lower boundary, but increases away from the boundary. Hence the velocity of the water film varies significantly across its thickness. Very large particles are virtually unaffected by the water flow, and very small light particles are carried away in suspension.

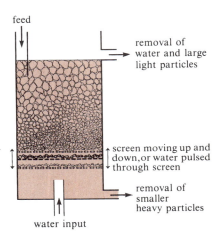

Figure 64　Jig.

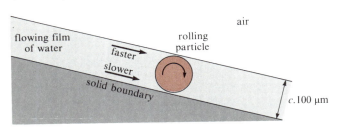

Figure 65　Particle rolling down an incline under the influence of a moving film of water.

Particles of approximately the same diameter as the thickness of the water film will roll down the incline. Lighter, rounder particles will roll further than heavier flatter ones. Figure 66 shows the idealized result of subjecting a range of particles to such a flow.

A *shaking table*[B] is a sloping surface shaken with a horizontal but asymmetrical action at right-angles to the flow of the water film. This reciprocating action causes any mineral particles fed onto the table to become separated into narrow bands, each consisting of one mineral type. Some tables are fitted with raised bars, which interrupt the water flow, and cause local turbulence and a jigging effect which aids separation. Shaking tables are used to separate mixtures of particles between about 50 μm and 1 mm in diameter (Figure 62).

shaking table

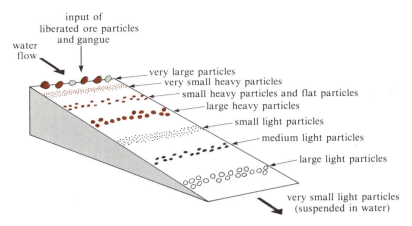

Figure 66 The sequence of particle separation in a flowing thin film of water.

input of
liberated ore particles
and gangue

water
flow

very large particles
very small heavy particles
small heavy particles and flat particles
large heavy particles
small light particles
medium light particles
large light particles

very small light particles
(suspended in water)

7.5.4 Magnetic separation

Different minerals can be magnetized to different degrees, and this variation in magnetic susceptibility forms the basis of a separation procedure, in a *magnetic separator*[B] (Figure 67). A stream of particles, ideally only one particle thick, is fed through a strong magnetic field. The more magnetic particles are deflected by the field, and can thus be collected separately from the less magnetic particles. Mineral particles larger than a few micrometres diameter can be separated by magnetic methods (Figure 62). Coarse-grained material is treated dry, but small particles tend to stick together if dry and hence are treated in a wet state.

7.5.5 Electrostatic separation

Electrostatic separation[B] is based on the mutual attraction of unlike *electrical charges*[A], and the mutual repulsion of like charges. The mineral particles are charged by an electrode as they are fed on to an earthed metal roller (Figure 68). Grains with low resistivity (good *conductivity*[A]) lose their *induced polarity*[A] and fall from the roller. Poorly conducting grains tend to retain their charges, stick to the roller for longer, and hence follow a different trajectory as they fall. Quartz

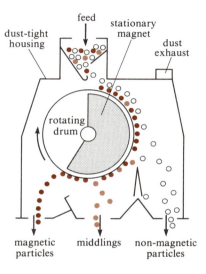

Figure 67 Magnetic separator.

feed
stationary
magnet
dust-tight
housing
dust
exhaust
rotating
drum
magnetic
particles
middlings
non-magnetic
particles

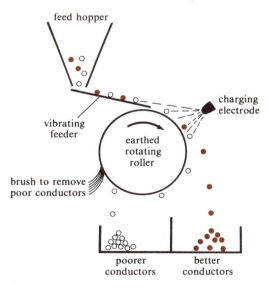

feed hopper

charging
electrode

vibrating
feeder

earthed
rotating
roller

brush to remove
poor conductors

poorer
conductors

better
conductors

Figure 68 Electrostatic separator.

grains are very poor conductors, stick quite firmly to the roller, and have to be removed by a brush. Electrostatic separation is often used to recover naturally liberated minerals such as ilmenite ($FeTiO_3$), zircon ($ZrSiO_4$) and cassiterite (SnO_2) from unconsolidated beach or alluvial placer deposits. The process works best on particles between 0.1 and 1 mm diameter (Figure 62).

7.5.6 Flotation

The principle of *flotation*[B] is to attach a mineral particle to an air bubble, so it is carried to the surface of a frothing liquid, whilst other minerals remain submerged in the liquid. Most minerals are *polar*[A], and thus have an electrically charged surface. Water molecules are also polar, and tend to stick to, or 'wet' uncontaminated mineral surfaces. Hydrocarbons, such as paraffin wax, are *non-polar*[A]. Their surfaces are not electrically charged, and they are not 'wettable'. By suitable chemical treatment, it is often possible to coat polar mineral particles selectively

flotation

61

with a thin film of a substance that will decrease their *'wettability'*[B]. Molecules of such compounds contain two structural features, a polar part, which adheres to the mineral surface, and a non-polar part, usually a hydrocarbon chain. A single mineral in a mixture of polar minerals in water, if selectively treated in this way, will become attached to an air-bubble when the mixture is aerated, rather than be 'wetted'. The selected mineral can be collected from the froth at the surface. Other reagents are often added to increase the persistence of the froth.

wettability

Flotation is widely used in the separation of sulphide minerals from silicates, in order to recover copper, zinc, lead, nickel, cobalt and molybdenum minerals. These sulphides often require separation not only from gangue, but also from each other. Figure 69 shows a single *flotation cell*[B]. In practice a series of cells is used, often to separate a series of different minerals. The maximum size of a particle that can be 'floated' depends on its density. Sulphide minerals with a density of about $5 \times 10^3 \, kg \, m^{-3}$ can be floated only if the particles are less than about $200 \, \mu m$ diameter. It is very difficult to achieve selectivity during a flotation process if the particles are very small. In practice the lower size limit is about $5-10 \, \mu m$ diameter (Figure 62).

flotation cell

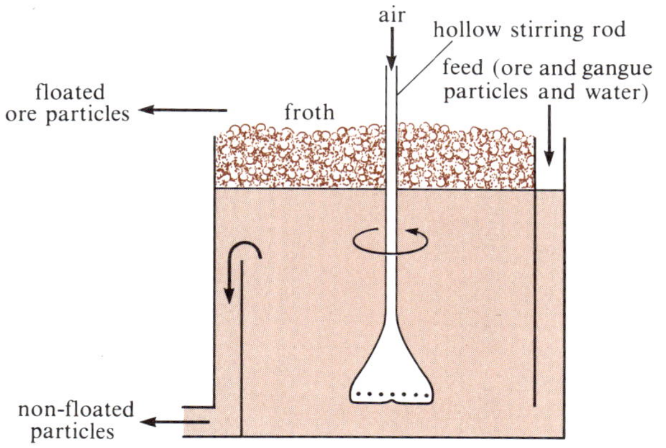

Figure 69 Flotation cell.

7.5.7 Leaching

In *leaching*[B] the minerals are attacked by suitable chemical reagents. Complete liberation is unnecessary, as the mineral grains need be only partly exposed to the reagent. If the material is coarse-grained, the reagent can percolate slowly through it, but fine-grained material must be continuously agitated to prevent the particles forming an impermeable mass. Most leaching processes dissolve more than the required element, so the solutions require additional treatment to recover the wanted metal.

leaching

The recovery of gold from quartz conglomerates containing submicroscopic fragments of the metal in their matrix provides a good example of leaching. Pulverized ore is treated with an aerated solution of sodium cyanide (NaCN). The gold is selectively dissolved as a complex cyanide:

$$4NaCN + 2Au + \tfrac{1}{2}O_2 + H_2O = 2Na[Au(CN)_2] + 2NaOH$$

The aeration also provides the agitation needed to maintain the ore particles in suspension. The gold-bearing solution is filtered, and zinc powder added, which precipitates metallic gold.

The ideal leaching solution should be cheap, selective to the desired mineral(s), and should not react with the containing vessel. Common examples are sulphuric acid (H_2SO_4), which is used to extract copper from fine-grained oxide ores, and sodium hydroxide (NaOH), which is used to extract alumina (Al_2O_3) from *bauxite*[A] ores.

7.6 Choosing a separation system

An ideal separation system is both cheap and efficient. Figure 70 summarizes the questions to be asked when choosing a separation system. In order to answer these questions, you will need to know the characteristics of the different processes, and the physical and chemical properties of the ore concerned. Figure 70 can then be used to choose a separation method or sequence of methods. If a sequence is chosen, the least expensive method should be used first, so that more expensive techniques handle a smaller throughput. You should also realize that Figure 70 is not an exhaustive list of separation techniques.

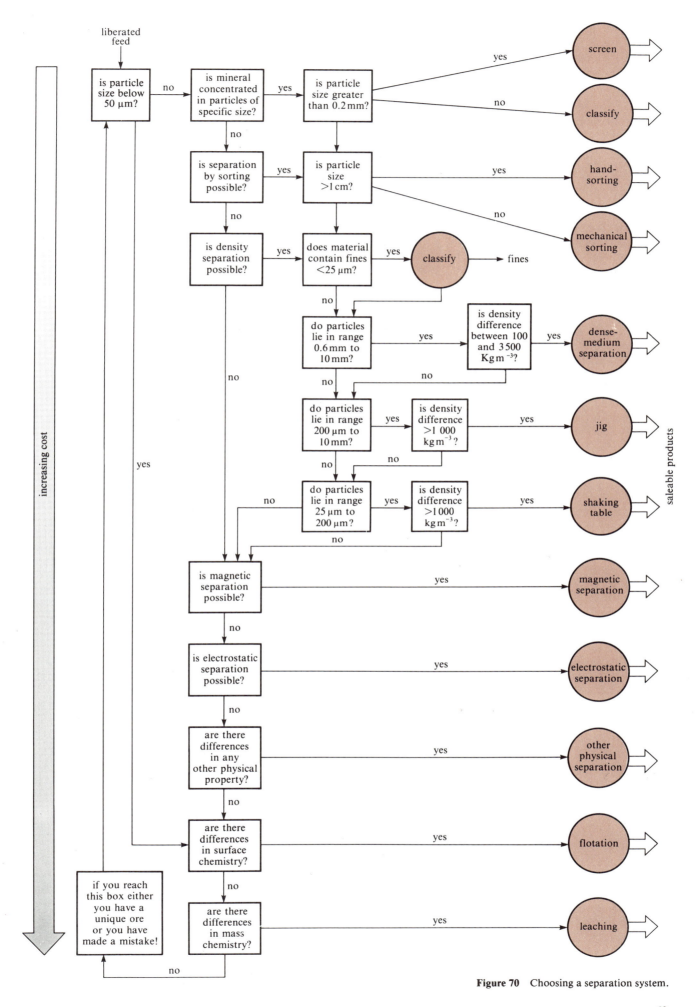

Figure 70 Choosing a separation system.

7.7 Drying procedures

The output from separation frequently contains large amounts of water. Removal of water to produce saleable concentrates uses one or more of three processes: thickening, filtering, and thermal drying.

During *thickening*[B] a suspension of solids is allowed to settle in a large container, so that a layer of clear liquid forms above a layer of solids. The solids slowly compact under the weight of overlying material. If fresh suspension is fed into the middle of the container, and the top and bottom layers are removed separately, then the thickening operation becomes a continuous process. The rate of compaction can be increased by slight agitation of the container, and the rate of settling can be enhanced by chemical *flocculating*[A] agents. Thickening can reduce water content from about 80 per cent to about 40 per cent.

thickening

Filtration[B] involves forcing a suspension of mineral particles through a porous material (usually closely woven cloth), the pores of which are too small to allow the particles to pass. Solids gradually build up on the filter, and because the water now has to pass through accumulated solid as well as through the filter, the rate of filtration decreases. The solids, which contain about 10 per cent of water, are removed at regular intervals (or continuously) in order to maintain an acceptable filtration rate.

filtration

If a moisture content below 10 per cent is required, then *thermal drying*[B] is used. In tropical areas the solid residues from filtration are dried by solar heating, but elsewhere coal, gas or oil heating is necessary. As thermal drying is costly, the water content is usually first reduced by thickening and filtration, so that less water has to be evaporated. Drying reduces the moisture content to almost zero, but fine-grained material may reabsorb up to 5 per cent of its own weight of atmospheric moisture on cooling.

thermal drying

7.8 Summary of Section 7

1   Ore is processed to produce a concentrate which is suitable for smelting and relatively cheap to transport, and no single process is optimal for more than one mineral assemblage. Qualitative and quantitative determinations of minerals in a deposit are necessary in order to design the best sequence of processes.

2   Mineral processing consists of three stages: liberation, separation, and drying. Each stage may have feedback loops to recycle insufficiently processed material. Flowsheets are used to represent these stages.

3   The higher the place value of an input to a processing plant, the more rigid is the constraint on where that plant can be sited.

4   Liberation involves breaking rocks to fragments smaller than the original grain sizes of the minerals, but does not in itself achieve any degree of separation of mineral from waste rock. Stressing a rock below its failure point is wasteful of energy, therefore the initial force applied should be large enough to fracture the rock.

5   Comminution can be considered in two stages: (i) crushing to about 1 cm diameter, (ii) grinding to smaller fragments. Comminution produces a range of particle sizes. The process is energy intensive, and the energy input increases exponentially with decreasing particle size. Over-reduction of size wastes energy, and may result in too fine a feed for subsequent separation processes. Hence particles already of sufficiently small size are continuously removed by screening or classifying.

6   Screening can be carried out wet or dry, and is a simple means of sizing particles larger than 1 mm diameter. Classification is used to size particles below 1 mm diameter, because the sinking velocity of such particles of similar densities varies with the square of their diameter. An appropriate upward current of water will then separate small slow-sinking particles from large fast-sinking ones.

7   Complete separation of minerals from gangue and from each other is rarely attained, and separation becomes increasingly complex and expensive as the particle size diminishes.

8   Very finely ground dense substances can be suspended in water to give 'dense media' liquids suitable for separating mineral particles by float–sink methods.

9   Jigs work by pulsation of water through a mineral bed, and separate small dense particles from large light ones. Shaking tables sort minerals into groups of distinct size and density by agitation in an inclined flow of water.

10   Minerals of different magnetic susceptibilities can be separated by magnetic fields, and minerals of different resistivities can be separated by exploiting the different rates at which they lose an induced polarity.

11   Differences in 'wettability' can be exploited to attach particles of a selected mineral to air bubbles, and remove it in the froth of an aerated mixture.

12   Separation using chemical solution methods (leaching) does not require complete liberation of minerals, and although costly, is often the only alternative if minerals are very finely disseminated.

13   Removal of excess water from mineral concentrates involves one or more of three successive stages: (i) thickening, a settling process that can be chemically assisted and reduces water content to 40 per cent; (ii) filtering, which reduces water content to about 10 per cent; (iii) thermal drying, which can remove all the water but is expensive. All three processes can be adapted to continuous operation.

14   Figures 62 and 70, respectively, summarize the scope and choice of techniques discussed in Section 7.

Objectives and SAQs for Section 7

Now that you have completed Section 7, you should be able to:

1   Define in your own words or recognize valid definitions of the terms and concepts introduced or developed in this Section and listed in Table B.

16   Explain the principles behind the methods of ore concentration, the factors governing choice of mineral processing techniques, and the problems of siting ore processing plants.

17   Explain the physical or chemical principles involved in, and the limitations of, crushing, grinding, screening, classification, mechanical sorting, float–sink separation, jigs, shaking tables, magnetic and electrostatic separation, flotation, leaching, thickening, filtering and drying. *(ITQs 22 and 23)*

18   Using quantitative data about the processes listed in Objective 17, perform simple calculations relating the nature of the feed to that of the product. *(ITQs 21–23)*

19   Using appropriate data about the minerals present in an ore, choose a liberation system and/or a separation system, and devise a flowsheet to represent those systems.

SAQ 18 *(Objectives 1 and 16)*   Which of the following statements are true?

(a) Liberation of minerals from basalt requires less size reduction than liberation of minerals from granite.

(b) Liberation of galena from quartz gangue decreases its place value.

(c) Separation of galena from quartz gangue decreases its place value.

(d) If a smelter is built on the mine site, there is no need to concentrate the ore.

(e) Choice of a separation method may be influenced by trace elements present in the ore minerals.

(f) Mineral processing should always take place on the mine site.

SAQ 19 *(Objectives 1, 18 and 19)*   A grinding process produces a range of particle diameters between 200 and 600 µm. The subsequent separation technique will work best on particles 200–300 µm diameter.

(a) Calculate the required velocity in the water column of a classifier to separate particles over 300 µm diameter from smaller particles (the particle density is 2.5×10^3 kg m^{-3}, the density of water is 1×10^3 kg m^{-3}, and the viscosity of water is 1×10^{-3} kg m^{-1} s^{-1}).

(b) Figure 71 shows an incompletely labelled flowsheet for the circuits associated with the classifier described in (a). Complete the flowsheet by inserting the missing labels (i)–(vii).

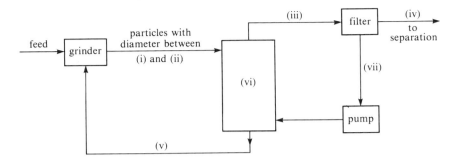

Figure 71 A flowsheet for classifier circuits.

SAQ 20 *(Objectives 1 and 19)* An orebody contains the minerals listed in Table 5, which also indicates their grain size, abundance and physical properties. Devise a flowsheet showing how, by successive operations, you would separate the useful minerals from the gangue and from each other.

Table 5 Properties of the minerals in a particular orebody

Mineral	Density/ $kg\,m^{-3}$	Hardness, Moh's scale	Magnetic properties	Grain diameter	Abundance (per cent)	Worth recovering from this ore?
chalcopyrite ($FeCuS_2$)	4 100	3.5	non-magnetic	150 µm	2	yes
molybdenite (MoS_2)	4 800	1.5	non-magnetic	150 µm	1	yes
pyrrhotite (FeS)	5 000	3.5	weakly magnetic	1–2 mm	7	no
feldspar ($KAlSi_3O_8$)	2 600	6	non-magnetic	more than 5 mm	30	no
quartz (SiO_2)	2 600	7	non-magnetic	more than 5 mm	60	no

SAQ 21 *(Objectives 1 and 17)* Explain why a mixture of large and small particles can be separated in a classifier, yet a similar mixture is deposited relatively unsorted in a thickening tank.

8 Metal smelting

Study comment This Section introduces the chemical principles of metal smelting. You have met the concepts of *oxidation*[A] and *reduction*[A] earlier in the Course, in the context of naturally occurring ores. Here we consider an 'unnatural' situation, the reduction of ore concentrates to parent metal, and see how the relative stabilities of ore minerals govern the theoretical amounts of energy required for smelting. The ITQs form essential steps in the reasoning, so please attempt them as you come to them, and read the answers before continuing. We also look at some practical aspects of smelting and the choice of method.

8.1 The stability of metal ores

Most metals occur naturally as stable compounds, such as oxides and sulphides. The formation of metal oxides and sulphides involves reactions that give off heat, i.e. the reactions are *exothermic*[A]:

$$\text{metal} + \text{oxygen} \rightarrow \text{metal oxide} + \text{(heat) energy}$$
$$\text{metal} + \text{sulphur} \rightarrow \text{metal sulphide} + \text{(heat) energy}$$

Smelting is the reverse of these reactions, being the separation of the metal from its combination in the oxide or sulphide ore mineral. It therefore involves reduction, and requires an input of energy, i.e. smelting reactions are *endothermic*[A]:

metal oxide + (heat) energy → metal + oxygen

metal sulphide + (heat) energy → metal + sulphur

The more stable the compound, the greater the energy liberated when it forms and hence the greater the energy required to recover the metal from that compound.

Although the central principle of metal smelting is reduction, each metal ore has its own chemical characteristics, and therefore requires a specific smelting process.

8.1.1 Metal oxides

Figure 72 is a graph which shows how the relative stabilities of a range of metal oxides vary with temperature. Such graphs are called *Ellingham diagrams*[B]. The horizontal scale is in degrees *Kelvin (K)*[A], a scale on which zero is equivalent to −273 °C, so that 300 K is about ordinary room temperature. The vertical scale is

Ellingham diagrams

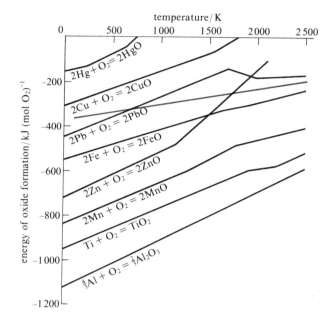

Figure 72 Relative stabilities of metal oxides. The identity of the brown line will be revealed later.

in energy units, and because each equation shown is standardized to one *mole*[A] of oxygen, the energy is expressed in kilojoules per mole of oxygen ($kJ (mol O_2)^{-1}$). Because all the reactions are of the form:

metal + oxygen → metal oxide + release of energy

the energy values on the graph are negative and represent the energy that has to be supplied to each system in order to decompose the oxide into metal and oxygen. Standardization to one mole of oxygen in each equation results in multiples or fractions for most of the metal components, for example 2 Zn and 4/3 Al.

The easiest way to understand how the graph works is to take an example. The simplest equation is $Ti + O_2 = TiO_2$. The line representing that equation on Figure 72 shows that at 500 K the formation of TiO_2 gives out 850 kJ of energy per mole of oxygen used. So at 500 K it will take 850 kJ to decompose a mole of titanium oxide to oxygen and titanium metal. Such an input of energy would require an increase in the temperature to well over 2 500 K, so clearly this is not a realistic method of obtaining titanium.

All the oxides shown on Figure 72 are stable at ordinary temperatures, and the most stable, the one which would require most energy to decompose, is alumina, Al_2O_3.

   Which is the least stable?

The one at the top of Figure 72, mercuric oxide (HgO). This would require zero energy to decompose into mercury and oxygen at about 700 K, and would decompose spontaneously at higher temperatures.

ITQ 24 (a) Above what temperature would a mixture of metallic mercury and oxygen be stable?

(b) At temperatures above 2 000 K, which of the oxides shown on Figure 72 will decompose into metal and oxygen?

(c) Where do you think a reaction involving oxygen, gold, and its oxide would appear on Figure 72?

Each reaction shown on Figure 72 represents a theoretical closed system involving the metal, its oxide, and oxygen, reacting at a pressure of one atmosphere. Clearly it is unrealistic to heat a metal oxide in an atmosphere of pure oxygen in order to obtain the metal. In practice, some metals are obtained by heating the oxide in air. Because air contains about 20 per cent oxygen, the effective pressure of that oxygen is only one-fifth of an atmosphere.

ITQ 25 If mercuric oxide were heated in air at atmospheric pressure, would the temperature at which it decomposed be higher or lower than the answer to ITQ 24(a)?

Only a limited number of metallic oxides can be reduced to metals by simple heating at temperatures up to 2 000 K, even if the oxides are heated in air as opposed to oxygen.

How can we exploit the fact that different metals have different affinities for oxygen?

In general, the lower that a metal oxide appears on Figure 72 the greater is its relative stability, i.e. the greater is the affinity of the metal for oxygen. This means that a metal will reduce the oxides of metals appearing above it.

ITQ 26 From Figure 72, can you predict the result of heating a mixture of copper oxide and iron filings?

Reduction of a metal oxide by another metal is costly, and is used only to obtain certain rare metals such as antimony. It is technically possible, but uneconomic, to obtain (say) iron by heating iron oxide with aluminium metal. In order to displace oxygen from iron oxides, we need a cheap and effective reducing agent. Carbon, hydrogen, and methane are good reducing agents, and carbon (as coke or charcoal) is cheap. The heating of metal oxides with coke forms the basis of many smelting processes. In terms of Figure 72, any metal oxide less stable than carbon monoxide will be converted into metal if heated in the presence of carbon. So, how stable is carbon monoxide?

The energy change involved in the reaction

$$2C + O_2 = 2CO$$

is about $-275 \, \text{kJ} \, (\text{mol} \, O_2)^{-1}$ at room temperature, (300 K) and about $-670 \, \text{kJ}$ $(\text{mol} \, O_2)^{-1}$ at 2 500 K, and is linear between these two points. Plot this information on to Figure 72, then try the following ITQ.

ITQ 27 (a) How does the stability of carbon monoxide vary with temperature, and how does this behaviour compare with that of a metal oxide?

(b) What are the implications of your answer to (a) with respect to the temperatures required to reduce metal oxides in the presence of carbon?

(c) At 2 000 K, which will be the more stable, CO or Al_2O_3? How feasible is it to attempt the reduction of alumina by heating it with carbon?

The answer to ITQ 27(c) is an example of the limitations on the heating of oxides with reducing agents. We shall return to the problem of aluminium smelting in Section 8.2.

8.1.2 Metal sulphides

In Part I you saw that many metals occur as sulphides rather than as oxides. Ellingham diagrams can be drawn for metal sulphides in the same way as for oxides (Figure 73). The reaction

$$2O_2 + S_2 = 2SO_2$$

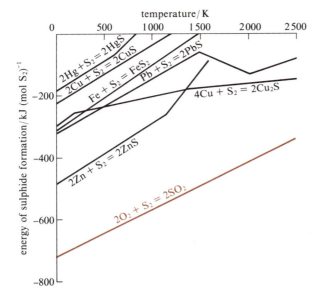

Figure 73 Relative stabilities of metal sulphides and sulphur dioxide.

has also been plotted on Figure 73, and shows SO_2 as more stable than FeS_2. Hence sulphur combines with oxygen more readily than with iron. Heating of FeS_2 in air leads to the reaction

$$2FeS_2 + 5O_2 = 2FeO + 4SO_2$$

In other words, iron *oxide* is produced, not metallic iron, and yet a comparison of SO_2 stability (Figure 73) with FeO stability (Figure 72) at first sight shows SO_2 as more stable than FeO.

Can you account for this apparent paradox?

The SO_2 reaction on Figure 73 involves *two* moles of oxygen, whereas the FeO reaction on Figure 72 involves only one. For a direct comparison, then, we need to express the SO_2 reaction as

$$O_2 + \tfrac{1}{2}S_2 = SO_2$$

This is shown as the brown line on Figure 72, and you can see that FeO is *more* stable than SO_2. This explains why the heating of FeS_2 in air produces iron oxide rather than metallic iron. In the past, FeS_2 was occasionally smelted for iron, but it is not normally an economic source of iron as iron oxides are readily available. Most metal sulphides are less stable than the corresponding oxides, and hence are easily oxidized further. If such a metal sulphide is heated in air, the sulphur present is oxidized to sulphur dioxide. Thus, ironically, oxidation of the sulphur is the first step in the reduction of many metal sulphide ores. However, if the metal oxide concerned is less stable than sulphur dioxide, the metal sulphide will be reduced directly to metal.

Two examples of metal sulphide smelting are:

1 Cinnabar (HgS) when heated in air gives the reaction

$$HgS + O_2 = Hg + SO_2$$

because as Figures 72 and 73 show, both HgS and HgO are less stable than SO_2 at 700 K. In fact, because mercury boils at 630 K it is released as a gas, and is condensed in water-cooled tubes before purification.

2 Sphalerite (ZnS) when heated in air gives the reaction

$$2ZnS + 3O_2 = 2ZnO + 2SO_2$$

This is because ZnS is less stable than SO_2 (Figure 73) whereas ZnO is more stable than SO_2 (Figure 72). Hence zinc oxide is formed, and must be reduced to metallic zinc by heating with powdered coal:

$$ZnO + C = Zn + CO$$

Because sulphide smelting can cause atmospheric pollution by sulphur dioxide, new smelters are built in areas where the environmental effects are least damaging, and often incorporate expensive anti-pollution systems in their design.

8.1.3 Other metallic compounds

Most metals in everyday use occur as oxide ores, or as compounds such as sulphides, carbonates or hydroxides which are easily converted into oxides. Most metal silicates are more stable than the corresponding oxides, and are more difficult and expensive to reduce. Consider the extraction of zirconium. Zircon ($ZrSiO_4$) is treated at high temperature with carbon. The product is chlorinated, and the resultant zirconium tetrachloride reduced by molten magnesium metal under a protective atmosphere of helium. Even then, the resultant metal needs to be refined by electrolysis.

8.2 Smelting practice

The stability of an ore mineral is not the only criterion governing the choice of smelting process. The reactions referred to on Figures 72 and 73 are all chemically possible, but some can be very slow, and are therefore impractical and uneconomic. Hence *reaction rates*[A] impose constraints on the choice of smelting process.

Various procedures are used to enhance reaction rates, such as:

1 Careful choice of temperature and pressure;

2 Use of a *catalyst*[A];

3 Prompt removal of products, so that *chemical equilibrium*[A] between reactants and products is not attained (*Le Chatelier's principle*[A]);

4 Removal of contaminants as early as possible in the process.

Even a high grade concentrate will contain some gangue. Some can be removed as a silicate *slag*[B] by adding a *flux*[A] such as calcium carbonate. Most slag results from reactions of the type:

slag

$$CaCO_3 + SiO_2 = CaSiO_3 + CO_2$$
$$\text{flux} \quad \text{in gangue} \quad \text{slag}$$

Low-density molten slag floats on the molten material accumulating at the bottom of the furnace and can be poured off.

As we saw in Section 8.1, some metals form very stable oxides which cannot economically be smelted by heating with a reducing agent. Such metals are obtained by electrolytic smelting methods. Aluminium is a typical example. It is smelted by dissolving alumina (Al_2O_3) in a molten mixture of cryolite (a mineral with formula Na_3AlF_6) and fluorite (CaF_2), and passing an electric current through the solution. Carbon electrodes are used. Aluminium collects at the *cathode*[A]. The *anode*[A] is consumed by the reaction:

$$2Al_2O_3 + 3C = 4Al + 3CO_2$$

As you will appreciate, decomposition of such a stable oxide at high temperature (the cryolite–fluorite mixture melts at 1 220 K) requires a very large input of energy.

8.3 Summary of Section 8

1 As the relative stability of a metal ore increases, so does the energy required to reduce it to metal.

2 Metals can be divided into six groups, according to the degree of difficulty of recovering parent metal from ore. Some metals are recovered from more than one ore, and hence occur in more than one group.

(i) Metals which occur in their native state and require no smelting (gold, silver, platinum, copper).

(ii) Metals easily recovered by roasting their ores in air (mercury, silver, copper).

(iii) Metals relatively easily obtained by reduction of their oxides (iron, manganese, chromium, tin).

(iv) Metals with ores that are easily converted into oxides, which are then reduced (iron, copper, lead, zinc, antimony, nickel, molybdenum).

(v) Metals with ores that are not easily smelted, and therefore require complex extractive processes (zirconium, niobium, tantalum, tungsten, titanium).

(vi) Metals that form very stable oxides, and are not amenable to extraction by simple heating with reducing agents. Such metals are usually extracted by electrolytic processes (sodium, potassium, beryllium, lithium, magnesium, calcium, aluminium).

3 Optimum smelting conditions are specific to each ore concentrate. Temperature and pressure are carefully controlled, catalysts added, and products removed, so that reaction rates and metal yields are maximized.

4 Contaminants that inhibit smelting processes or are undesirable in the product are combined with a flux to form a removable slag.

Objectives and SAQs for Section 8

Now that you have completed Section 8 you should be able to:

1 Define in your own words, or recognize valid definitions of the terms and concepts introduced or developed in this Section and listed in Table B.

20 Explain the physical and chemical principles of metal smelting, and write simple chemical equations describing the reactions involved. *(ITQs 24–27)*

21 Relate the stabilities of metallic ore to an appropriate choice of smelting techniques *(ITQ 27)*

SAQ 22 *(Objectives 1, 20 and 21)* Which of the following statements are true at temperatures of 1 000 K?

(a) Oxygen has a greater affinity for manganese than for zinc.

(b) Sulphur has a greater affinity for zinc than for copper.

(c) Oxygen has a greater affinity for zinc than for sulphur.

(d) Sulphur has a greater affinity for oxygen than for zinc.

(e) Zinc has a greater affinity for oxygen than for sulphur.

(f) Manganese metal will be displaced from manganese oxide if the oxide is heated with powdered aluminium metal.

(g) A mixture of carbon and iron oxide (FeO) will not yield iron metal.

SAQ 23 *(Objectives 1 and 20)* Which of the following physical properties would be desirable in a slag? (i) A high melting point; (ii) a low melting point; (iii) a high viscosity; (iv) a low viscosity; (v) a high density; (vi) a low density.

Objectives and Table B for Block 3 Part II

When you have completed Block 3 Part II you should be able to:

1   Define in your own words or recognize valid definitions of the terms and concepts introduced or developed in Block 3 Part II and listed in Table B.

2   Discuss the factors underlying decisions to initiate exploration, select an exploration objective, assess the risks in exploration, and assign a budget to exploration. *(ITQ 1; SAQs 1 and 2)*

3   Use information about different rock types, their regional setting and relationships to one another to suggest sites of potential mineralization. *(ITQs 2, 3 and 4; SAQ 3)*

4   Recognize and delineate simple geological features, such as faults, igneous intrusions and banded rocks, on remotely sensed images. *(ITQ 5; SAQ 4)*

5   Select combinations of multispectral data that most clearly distinguish different surface materials. *(ITQ 6; SAQ 5)*

6   Using data on the distribution of an element over an area, suggest a threshold, identify anomalies and describe them, and suggest further sampling patterns. *(ITQs 7, 8 and 12; SAQ 8)*

7   Outline the characteristics of different primary and secondary dispersion patterns and account for them in terms of the dispersion processes involved. *(ITQs 9 and 10; SAQs 6 and 9)*

8   Interpret examples of data from geochemical surveys of: (a) bedrock; (b) soil; (c) natural water; (d) drainage sediments; (e) vegetation. *(ITQ 11; SAQ 7)*

9   Describe the principles behind the interpretation of anomalies defined by gravity, magnetic and electrical surveys. *(ITQ 13; SAQ 13)*

10   Select the type of geophysical survey most useful for detecting different kinds of mineral deposit. *(SAQ 10)*

11   Using data from a geophysical survey, identify anomalies and make qualitative estimates of the shape and depth of the bodies responsible for them. *(ITQs 14 and 15; SAQ 12)*

12   Estimate the costs of evaluation programmes and compare them with those relating to exploration and mine development. *(ITQ 16)*

13   Use drill core data to assess the depth, shape and value of mineral deposits. *(SAQ 14)*

14   Recognize and evaluate the various economic, geological, technical and other factors controlling the viability of surface and underground mines. *(ITQs 17–20; SAQs 15 and 16)*

15   Identify the geological conditions suitable for surface and underground mining and their relationship to operating methods. *(SAQ 17)*

16   Explain the principles behind the methods of ore concentration, the factors governing choice of mineral processing techniques, and the problems of siting ore processing plants. *(SAQ 18)*

17   Explain the physical or chemical principles involved in, and the limitations of, crushing, grinding, screening, classification, mechanical sorting, float–sink separation, jigs, shaking tables, magnetic and electrostatic separation, flotation, leaching, thickening, filtering and drying. *(ITQs 22 and 23; SAQ 21)*

18   Using quantitative data about the processes listed in Objective 17, perform simple calculations relating the nature of the feed to that of the product. *(ITQs 21–23; SAQ 19)*

19   Using appropriate data about the minerals present in an ore, choose a liberation system and/or a separation system, and devise a flowsheet to represent those systems. *(SAQs 19 and 20)*

20   Explain the physical and chemical principles of metal smelting, and write simple chemical equations describing the reactions involved. *(ITQs 24–27; SAQs 22 and 23)*

21   Relate the stabilities of metallic ores to an appropriate choice of smelting techniques. *(ITQ 27; SAQ 22)*

Table B Terms or concepts introduced or developed in Block 3 Part II

Term or concept	Page	Term or concept	Page	Term or concept	Page
alluvial mining	45	flotation	61	open pit	43
background	16	flotation cell	62	open-pit mining	45
ball mill	57	flowsheets	54	ore passes	51
bench	45	footwall	48	orientation survey	21
biogeochemical survey	21	geobotanical survey	20	persistence	20
blocking out	37	geochemical anomaly	16	primary dispersion	18
Bond work index	56	geochemical dispersion	18	radioactivity survey	25
caving	51	geophysical anomaly	24	raise	51
classifier	58	gravity anomaly	26	reduction ratio	56
closed circuit	56	grinding	56	reflectance	13
comminution	56	hanging wall	48	resistance	30
concentrates	54	horizontal profiling	31	resistivity	30
cone crusher	56	indicator plants	20	rock associations	10
contrast	16	induced polarization (IP) survey	33	rod mill	57
crosscut	50	jaw crusher	56	room and pillar mining	49
crushing	56	jig	60	secondary dispersion	18
cut-and-fill stoping	51	leaching	62	separation	54
cut-off point	20	liberation	54	shaft	48
cut-off value	40	line electrode method	31	shaking table	60
cyclone	58	lithogeochemical survey	18	shrinkage stoping	51
decline	48	locally enriched deposits	40	slag	70
dense media	59	longwall mining	50	smelting	54
drainage sediment survey	20	magnetic relief	29	soil survey	19
draw point	51	magnetic separation	61	spontaneous polarization (SP) survey	32
drill core	35	magnetic susceptibility	28	stoping	51
drift	48	maximum economic stripping ratio	44	stressed vegetation	21
electromagnetic (EM) survey	33	mechanical sorter	59	stripping ratio	43
electrostatic separation	61	mining cut-off	40	sub-level stoping	51
Ellingham diagrams	67	nanotesla	28	thermal drying	64
exploration objective	7	natural water survey	20	thickening	64
exploration risk	7	ohms	30	threshold	16
false-colour image	14	ohm metres	30	unit cost	39
filtration	64	open-cast mining	47	unit value	40
fixed costs	39	open circuit	56	vertical sounding	31
float–sink separation	59			wettability	62

ITQ answers and comments

ITQ 1 Figure 1 suggests that three $(6-3)$ deposits of 10^7 tonnes, and six $(9-3)$ deposits of 10^5 tonnes, remain to be discovered in the mining district. If the number of known deposits of 10^7 tonnes was eight or nine then this might lead to speculation that more deposits of 10^8 tonnes were present. However, such a method is crude without more information on the types of deposit present in the district.

ITQ 2 (i) Porphyry copper deposits are generally associated with volcanic arcs, though relics of such terrains may be preserved in cratons.

(ii) By their very nature as chemical precipitates, stratiform deposits are expected to have formed in sedimentary basins.

(iii) Magmatic segregation deposits of chromium are found in gabbroic bodies, so would be most likely to occur in ancient fragments of oceanic crust simply because the latter commonly contain such bodies. However, the world's most productive chromium mines are located in gabbroic intrusions — the Great Dyke of Zimbabwe and the Bushvelt Complex of South Africa — emplaced in cratons.

(iv) Banded ironstones formed under conditions that only existed at the Earth's surface before about 1 500 Ma ago. Thus the search for such deposits must, by definition, be focused on cratons.

ITQ 3 (a) Table 6 in Part I showed you that high concentrations of tin and uranium are associated with granite, so an absence of such rocks would make a search for ore deposits of these metals not worthwhile. Chromium is associated with gabbro and peridotite, so again, an absence of these would make the occurrence of chromium ores unlikely.

(b) Both chromite and cassiterite (the tin ore, SnO_2) are hard and dense and resistant oxides, and they might well occur among the gravels of modern stream channels as *placer deposits*A, provided of course that the primary source rocks were in the vicinity to be eroded. Uranium is soluble in oxidizing groundwaters, but is precipitated in reducing conditions such as might be expected in the vicinity of plant remains (cf. Part I, Figure 51). So — again provided that source rocks were available — uranium would be sought in the sandstones.

ITQ 4 (i) Being dense, massive sulphide segregation deposits tend to accumulate at the base of intrusions. (ii) Pegmatites form from residual watery fluids in granites, which separate and crystallize in the fissured upper zone of intrusions. (iii) Again, hydrothermal deposits tend to form *above* a source of heat, fluid and metals, in suitable host rocks such as limestones. However, those of lead and zinc crystallize at relatively low temperature and so form at a distance from the actual intrusion contact. (iv) Secondary enrichment requires the penetration of surface water into the primary ore, so such deposits would be located beneath the exposed upper surface of the intrusion.

ITQ 5 (i) Figure 74 shows boundaries that are easily interpreted from Figure 6 and which cut the sedimentary layers. They outline light-coloured rocks devoid of banding, which are granite intrusions. (ii) We have separated them into elliptical granites and narrow quartz veins. (iii) The roughly straight NNW–SSE features cut both sediments and granites and are therefore younger than both. They could be faults, but major boundaries do not appear to have been displaced, and they are probably narrow igneous dykes which erosion has picked out.

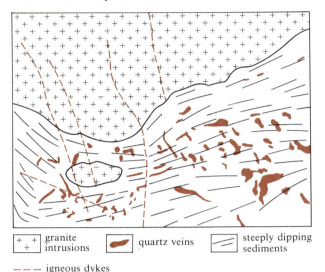

| + + | granite intrusions | quartz veins | steeply dipping sediments |

- - - igneous dykes

Figure 74 Answer to ITQ 5. Geological interpretation of Figure 6.

ITQ 6 All the combinations show large differences between grass and fir because of the sharp change in reflectance on going from visible to infrared wavelengths. The ratio of band 7 to band 5 gives the greatest difference. For andesite and shale, which show more gradual increases in reflectance on going from visible to infrared wavelengths, only the ratio of band 7 to band 4 gives a clear distinction. Ratios between bands are useful because they indicate the *gradients* of reflectance curves, and as Figure 7 shows, these differ markedly for different materials. The completed table is as follows.

Band combination	Grass	Fir	Percentage difference	Andesite	Shale	Percentage difference
4 + 7	120	65	45	37	34	8
7 − 4	80	15	81	10	12	17
7:4	5	2	60	1.3	2	35
7:5	31	2	94	1.1	1.2	8
4:7	0.2	0.3	33	0.1	0.08	20

ITQ 7 (a) The background range is about 20–80 p.p.m. (b) The clearly anomalous values range from 80 to 300 p.p.m. (c) For an average background of 50 p.p.m. the contrast is about 300:50 or 6:1 (d) The threshold is at about 100 p.p.m.

ITQ 8 Figure 75 shows the areas of further exploration outlined for a threshold of 45 p.p.m. (brown) and one of 250 p.p.m. (black). In the first case a very large area is defined as an anomaly; in the second the anomaly, as defined, is small. So with a threshold of 45 p.p.m., much time and money may be used in the examination of insignificant areas, whereas with a threshold of 250 p.p.m. some ore deposits may be missed.

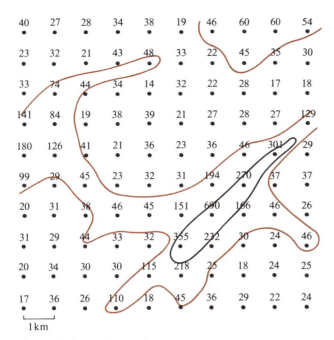

Figure 75 Answer to ITQ 8.

(Part I, Section 4.4). If they are soft then they will be rapidly broken to small grains, ensuring that they are transported away before an anomaly can form, irrespective of their density. The most likely contributor to drainage sediment anomalies is therefore a hard mineral of high density.

ITQ 11 Stream system (b) will show the greatest persistence. Stream system (a) includes a major river flowing from unmineralized areas which will dilute any anomaly. Stream system (c) includes a lake in which anomalous ore grains will be deposited or dissolved metals may be precipitated, leading to reduced persistence.

ITQ 12 Figure 76a shows the completed histogram and the frequency distribution curve. The curve shows that there are two distinct sample populations in the area, the majority defining the background, peaking at between 20 and 40 p.p.m. copper, and a less common population peaking at 125−160 p.p.m. copper which form the anomalous soil samples. The boundary between the two curves is at about 70 p.p.m., which is the best threshold on which to base further exploration.

Figure 76b shows the areas of significant anomaly and hence the targets for further more detailed exploration.

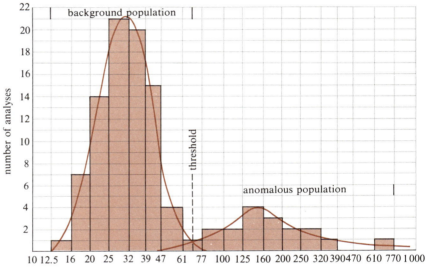

ITQ 9 (a) Magmatic segregation depends on the formation of immiscible sulphide droplets in a gabbroic or peridotite magma. Elements with an affinity for sulphur, such as nickel, enter the sulphide melt preferentially and become enriched. The dense sulphide droplets accumulate at the base of an igneous mass. The mass is sharply divided into nickel-poor and nickel-rich parts and dispersion is at a minimum.

(b) Porphyry copper deposits are produced by metal-bearing fluids moving along tiny cracks in a granodiorite igneous mass. There is no sharp division between copper-poor and copper-rich zones so dispersion is very broad.

(c) Pegmatites form from magmatic fluids in sharply defined veins[A] and so dispersion is narrow.

(d) Placer deposits are formed by winnowing of light, soft or soluble minerals from dense, hard, insoluble minerals over protracted periods usually in flowing water. Dispersion may be broad along the length of flow but very narrow across the dispersing channel.

Porphyry copper deposits therefore show the broadest primary dispersion.

ITQ 10 If they have the same grain size as the predominant quartz and common rock particles in a stream, dense ore minerals will not be transported as far as the bulk of the sediment and will form an anomaly similar to a placer deposit

Figure 76 Answers to ITQ 12.

ITQ 13 Figure 77 shows a completed Figure 22. Note that above the dipole the lines of induced magnetic flux are opposed to those of the Earth, consequently the resultant is a negative total field anomaly over the dipole.

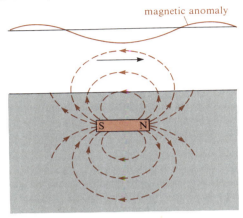

Figure 77 Answer to ITQ 13.

ITQ 14 Figure 78 shows the areas of low and high resistivity, where lines of equal potential become more widely separated and bunched-up, respectively.

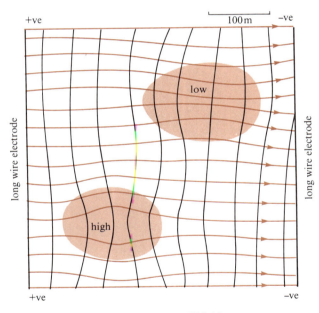

Figure 78 Answer to ITQ 14.

ITQ 15 The lines of equal potential bunch up and assume extreme values near the intersection of grid lines G–G and 21. This suggests that this area overlies an electrically polarized body probably resulting from oxidation of sulphides.

ITQ 16 The amount of diamond drill coring required to explore 1 km² is $10 \times 10 = 100$ holes, which at 200 metres per hole gives 20 000 metres. At 40 US dollars per metre this would cost 800 000 US dollars for drilling alone. Remote sensing plus EM surveys and interpretation of 900 km² would cost about 650 and 60 000 US dollars, respectively. Geological mapping and soil surveys of 100 km² would cost 60 000 and 75 000 US dollars, respectively.

ITQ 17 (a) The bulk of the reserves in a confined deposit are of high grade ore which is strongly contrasted to the host rock, so that the reserves rise rapidly as the grade decreases to a limiting value and then do not increase much until the grade falls to crustal abundance. Consequently, if the cut-off grade is in the large range between crustal abundance and the limiting value, the reserves stay roughly constant. Therefore, provided the cut-off grade is less than the limiting value in the deposit, the reserves bear little relation to the metal price or the unit cost.

(b) Compared with the cut-off grade indicated on Figure 35, that for which the reserves fall to half is roughly three to four times higher. Therefore the price may drop to a quarter of the level indicated by the diagram or the costs may rise by four times before the reserves fall below half their indicated tonnage. Compared with a dispersed deposit, a confined deposit has constant reserves even when the metal price and the unit costs fluctuate greatly.

ITQ 18 Figure 37 shows that the upper enriched part of the deposit is a confined high grade ore, whereas the lower part is a dispersed orebody. The grade–tonnage curve will therefore be roughly a combination of Figures 34 and 35, as shown in Figure 79. With constant metal price and high unit costs, the cut-off grade (A) may be so high that only the enriched part constitutes reserves. Falling unit costs eventually cause the cut-off grade to lie within the range of grades of the dispersed, deeper parts (B) and the reserves may suddenly become up to three or four times greater. This is the background for the progressive mining of porphyry copper deposits where initially high unit costs, dominated by fixed costs, meant that only secondarily enriched zones could be mined. When the capital cost of permanent equipment in the mine was eventually paid for, unit costs decreased, so that lower grade ores became profitable.

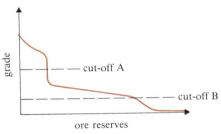

Figure 79 Answer to ITQ 18.

ITQ 19 The cross-sections of the pit at successive stages are shown in Figure 80. The stripping ratios are calculated as follows:

$$\text{level 2 stripping ratio} = \frac{212 \times 2600}{324 \times 4500} = 0.38$$

$$\text{level 3 stripping ratio} = \frac{380 \times 2600}{290 \times 4500} = 0.76$$

$$\text{level 4 stripping ratio} = \frac{540 \times 2600}{260 \times 4500} = 1.20$$

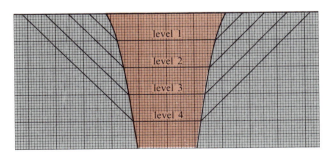

Figure 80 Answer to ITQ 19.

You may have counted slightly different numbers of squares, but the increase in stripping ratio with depth should be similar.

ITQ 20 The overall stripping ratio in the existing pit containing A is $572/790 = 0.72$. The stripping ratio for the remainder, B, is $837/308 = 2.7$.

If ore with a grade of 3.0 per cent metal is just profitable with a stripping ratio of 0.76 the ore in B must have a grade of

$$\frac{3.0 \times 2.7}{0.72} = 11.25 \text{ per cent}$$

ITQ 21 (a) Cassiterite contains

$$\frac{119}{119 + (2 \times 16)} \times 100 \text{ per cent tin}$$

$$= 79 \text{ per cent tin}$$

(b) The ore contains 1.5 per cent cassiterite, of which 79 per cent is tin, and 0.5 per cent stannite, of which 28 per cent is tin. So the total tin content of the ore is

(79 per cent of 1.5 per cent) + (28 per cent of 0.5 per cent)
= 1.18 per cent + 0.14 per cent
= 1.32 per cent tin

(c) There is 1.32 per cent tin in the ore, of which 1.18 per cent is in cassiterite. Assuming complete recovery of the cassiterite and its tin content (not achieved in practice) the tin recovered is $1.18/1.32$ of the total tin content, which is 89 per cent.

ITQ 22 (a) If $W_i = 12$ kWh per short ton and $F_{80} = 100\,\mu\text{m}$, then a ten-fold reduction in size gives $P_{80} = 10\,\mu\text{m}$. So

$$W = 10 \times 12 \times \left[\frac{1}{\sqrt{10}} - \frac{1}{\sqrt{100}} \right]$$
$$= 120 \times (0.32 - 0.1)$$
$$= 26 \text{ kWh per short ton}$$

(b) If $W_i = 12$ kWh per short ton and $F_{80} = 1\,000\,\mu\text{m}$, then a ten-fold reduction in size gives $P_{80} = 100\,\mu\text{m}$. So

$$W = 10 \times 12 \times \left[\frac{1}{\sqrt{100}} - \frac{1}{\sqrt{1000}} \right]$$
$$= 120 \times (0.1 - 0.032)$$
$$= 8 \text{ kWh per short ton}$$

These results show that considerably more energy is needed to reduce the size of small particles than to reduce the size of large particles. In fact, the increase is exponential.

ITQ 23 (a) Equation 4 shows that the settling rate varies directly with the density difference between particle and fluid, but with the *square* of the particle diameter. Hence size is more important than density in determining settling rates, and very small dense particles (ferrosilicon) will settle more slowly than larger less dense particles (mineral fraction).

(b) The mineral particles settle with

$$V_m = \frac{(3.5 - 3) \times 10^3 \times (1 \times 10^{-3})^2 \times g_x}{18 \eta_x}$$

The ferrosilicon particles settle with

$$V_f = \frac{(6.5 - 3) \times 10^3 \times (100 \times 10^{-6})^2 \times g_x}{18 \eta_x}$$

The settling forces, g_x, and the viscosities, η_x, are unknown but common to both equations, so the *ratio* of the two velocities is

$$\frac{V_m}{V_f} = \frac{(3.5 - 3) \times (1 \times 10^{-3})^2}{(6.5 - 3) \times (100 \times 10^{-6})^2}$$
$$= \frac{0.5 \times 10^{-6}}{3.5 \times 10^{-8}}$$
$$= 14:1$$

This means the mineral particles (1 mm diameter) sink 14 times faster than the ferrosilicon particles (100 μm diameter).

ITQ 24 (a) Above about 700 K.

(b) HgO and CuO.

(c) Gold is found as the native metal, even when in the form of finely divided grains in an oxidizing environment. Clearly the oxide is not readily formed, and the reaction line would be expected to appear above that for mercury.

ITQ 25 Lower. Because the effective pressure of the oxygen is only one-fifth of an atmosphere, the lower oxygen pressure will favour decomposition into mercury and oxygen (*Le Chatelier's Principle*[A]).

ITQ 26 Figure 72 shows iron oxides as more stable than copper oxides. Therefore one would expect iron to displace copper from combination with oxygen, provided enough heat was available to initiate the reaction:

$$Fe + CuO = FeO + Cu$$

ITQ 27 (a) See Figure 81. The stability of CO increases with temperature, in contrast to the stabilities of metal oxides which decrease with temperature.

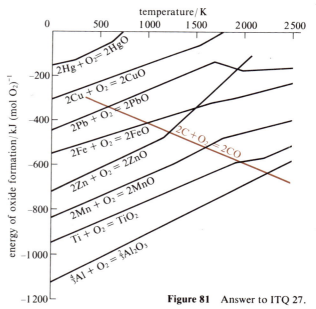

Figure 81 Answer to ITQ 27.

(b) The implication is that many metallic oxides can be reduced by carbon, because heating the oxide with carbon, to a temperature above the point where the CO stability slope crosses that of the metal oxide, will produce CO and the appropriate metal.

(c) At 2 000 K, Al_2O_3 is more stable than CO. Temperatures above 2 300 K would be necessary to reach a point where CO was the more stable, so although technically possible, reduction of alumina by carbon requires too high a temperature to be economically feasible.

SAQ answers and comments

SAQ 1 (a) In 1974 the economic climate was erratic for copper but stable for tin, for which demand was inelastic. Tin would have been a better target than copper.

(b) There is little likelihood that accessible plant will be near any new discovery, so a large deposit that can support a complete mining and processing operation is a better bet than some small deposits that may be difficult to link up. The risk of not finding anything is high so it is better to aim for a large deposit for this reason too.

(c) Because the area is politically stable there is no need to plan for a brief but highly profitable high grade operation, although such a discovery would not be dismissed.

(d) The exploration risk being high, only a low proportion of returns can be assigned safely to exploration. Yet another reason for a large target deposit is that in such an area, costs of any kind of exploration will be higher than usual.

(e) Given the present air of financial uncertainty and the undeveloped nature of the area, the primary objective would be selection of favourable areas to hold awaiting further developments. In such an unknown area the chance of making an exciting discovery at some time is higher than in an area that has been well explored.

SAQ 2 For an ore containing 10 per cent copper a single deposit containing more than a million tonnes of metal must contain at least 10 million tonnes of ore. For an ore containing 1 per cent copper the tonnage is 100 million tonnes, and for 0.1 per cent copper it is 1000 million tonnes. Figure 82 shows clearly that a million tonnes of copper metal can only be expected in a single deposit that appears to the right of a line joining these three points. Only a stratiform or a porphyry copper deposit will fit the bill. If there is a need to focus on a high grade deposit which can be exploited quickly, as in an insecure country, then the choice is cut down to stratiform deposits.

SAQ 4 (i) and (ii) The east—west dark lines stop abruptly at the western-most of the curved features and are therefore older. The rocks through which the dark lines pass have a vague north—south set of linear features, which may be older than both sets of lines mentioned in the question. (iii) The east—west lines are basaltic dykes and the curved linear features are escarpments related to younger thick sedimentary beds that dip gently to the east.

SAQ 5 Sandstone has a much higher reflectance than andesite in bands 7 and 5, but lower reflectance in part of band 4. The sandstone has roughly equal brightnesses in bands 7 and 5. A false-colour image of sandstone would have equally high intensities of red and green and a low intensity of blue. In fact this gives a yellow hue. Andesite has roughly similar brightnesses in red, green and blue. Equal intensities of the three primary colours gives grey tones from white to black. Because all three are low, the andesite would appear dark grey. (An explanation of the way colours are produced by mixing red, green and blue light is given in AV6.)

SAQ 6 (i) Granites normally crystallize at temperatures around 700 °C so this is probably secondary dispersion; however, the fluids may be related to cooling of the intrusion in which case they may be regarded as primary dispersion agents. (ii) Secondary, because the deposit was formed by earlier igneous activity. (iii) Secondary. (iv) Secondary, though if the fragments are exploitable, this could be regarded as primary dispersion.

SAQ 7 Samples may be taken from groundwater by boreholes proceeding uphill. Soil surveys might reveal a soil anomaly (cf. Figure 10e). Foliage of deeply rooted trees on the slope may be analysed.

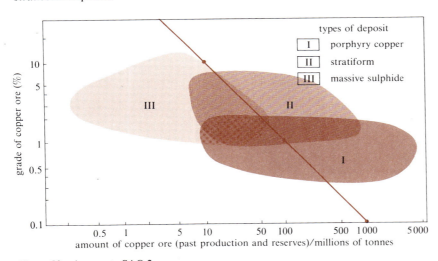

Figure 82 Answer to SAQ 2.

SAQ 3 (a) Pegmatite deposits are often concentrated at the boundaries of granite intrusions, and lithium is both an abundant trace element in granite magmas and also selectively concentrated in residual liquids as the latter crystallize.

(b) The quartz veins cutting the sediments are the most likely sites of hydrothermal lead—zinc mineralization (cf. Part I, Section 3.3).

SAQ 8 The background ranges from 15 to 77 p.p.m. copper with an average of about 32. The maximum anomaly is 690 p.p.m. copper, therefore the contrast is 690:32 or about 21.5:1.

SAQ 9 The minerals containing mercury, nickel and copper are soluble and soft, and only those for chromium and tin satisfy the requirements for a drainage sediment survey.

SAQ 10 (a) All three survey techniques would be capable of detecting massive copper–nickel sulphides because the ore minerals are dense and at high concentration, have low resistivity, and, in the case of pentlandite, have a high magnetic susceptibility compared with common rocks and minerals.

(b) Gravity surveys would reveal the presence of low density rock, but not its composition. Neither the electrical nor the magnetic properties of cassiterite favour its detection by geophysical means, especially as the veins are thin and deeply buried, thereby degrading the resolution of all techniques.

(c) Chromite and gabbro have similar resistivity, density and magnetic susceptibility. It would be almost impossible to detect the chromite by geophysical means alone unless it was associated with (valueless) magnetite or ilmenite which also segregated into bands. In that case magnetic anomalies over the gabbro may indicate possibly favourable areas.

SAQ 11 Spontaneous polarization surveys depend on the natural oxidation of sulphides by groundwater and the production of currents, whereas induced polarization uses pulses of current to set oxidation going in wet sulphides and detects the weak remaining potential when power is turned off.

SAQ 12 The main anomaly pattern is nearly circular and comprises both negative and positive regions. This is very similar to that associated with a buried dipole, similar to a magnetic sphere at high latitudes. However there is a marked east–west alignment in the anomaly pattern which may indicate a fault as well. A fault by itself would give only a linear anomaly pattern, whereas a horizontal sheet would give a uniformly high magnetic field.

SAQ 13 Figure 83 shows the arrangement of lines of flux above an isolated magnetic pole (the opposite pole would have no effect), and the appearance of the anomaly in cross-section. In plan it would comprise a central positive anomaly surrounded by a circular negative anomaly.

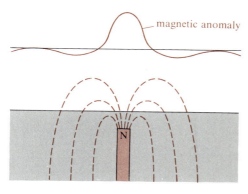

Figure 83 Answer to SAQ 13.

SAQ 14 Figure 84 shows a completed version of Figure 33.
(a) The orebody appears to be a tabular sheet that is nearly horizontal in the area covered by holes A, B, C, D, E, F and G. However, in holes H and I there is a marked increase in depth to the orebody, suggesting either a fault or a fold. The vertical plane shown in Figure 84 represents an inferred fault.

(b) The block is 500×500 metres in area, the average thickness of ore is $(150 + 100 + 100 + 20)/4 = 92.5$ metres, and its average grade is $(0.8 + 3.0 + 0.4 + 0.7)/4 = 1.23$ per cent. The tonnage of copper is $(500 \times 500 \times 92.5 \times 3.0 \times 1.23)/100 = 8.53 \times 10^5$ tonnes, which has a value of about £700 million.

(c) The most favourable area for further drilling is south of C and B where grades are highest and thickness seems to be increasing southwards.

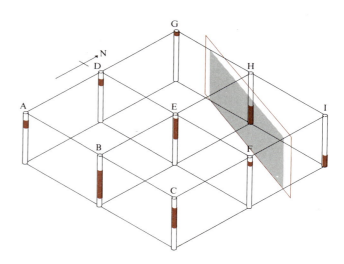

Figure 84 Answer to SAQ 14.

SAQ 15 The chemical precipitate (ii) is a dispersed deposit and its reserves are therefore extremely sensitive to changes in the cut-off grade. Deposit (i) is a confined, bonanza-type deposit and (iii) is partly confined and partly dispersed.

SAQ 16 (a) The stripping ratios for successive benches at 20, 40 and 60 metres are about 250/1100, 820/900 and 1400/700 which give 0.23, 0.91 and 2.00, respectively, so that open-pit mining of this deposit has a depth limit of 60 metres.

(b) The overall stripping ratio for the bottom 40 metres, from 60 down to 100 metres is roughly 4700/780 which is about 6.0. If the ore down to 60 metres has an average grade of x and is just profitable with a stripping ratio of 2.0, the ore in the bottom 40 metres must have a grade of at least:

$$\frac{6.0}{2.0}x = 3x$$

if the entire deposit is to be worked. (This exercise is analogous to the situation in the Pine Point lead–zinc orebodies (Part I, Section 3.3), and if the pipe-like deposit continued much deeper at high grade, underground mining would be required.)

SAQ 17 (a) Both the ore and the roof rock are strong enough to support workings, and room and pillar methods would be best, perhaps with backfilling to allow pillars to be removed at a late stage.

(b) The ore is strong enough for support but the walls would be unstable, therefore cut-and-fill stoping must be used to prevent inward collapse of walls.

(c) The deposit is ideal for caving methods; however, the ore is too strong to collapse and fragment under its own weight, therefore caving must be induced by drilling and blasting.

(d) The ore is thick but weak, and unsupported stoping would be dangerous. The safest technique would be shrinkage stoping.

SAQ 18 (a) False; basalt is finer grained than granite (Block 1) so would need to be reduced to smaller grain sizes to liberate the minerals.

(b) False; liberated material is still at the same grade as uncrushed ore.

(c) True; this process forms a concentrate, which may require drying before it is saleable.

(d) False; concentration processes are carried out not only to reduce transport costs, but also to produce an acceptable smelter feed.

(e) True; traces of other elements may alter the physical or chemical properties on which the separation procedures are based.

(f) False; although true in general, there are situations where the mined grade is rich enough to justify transport to a central processor serving several mines.

SAQ 19 (a) Using equation 4 we get

$$V = \frac{(2.5 - 1) \times 10^3 \times (300 \times 10^{-6})^2 \times 9.8}{18 \times 1 \times 10^{-3}}$$
$$= 7.35 \times 10^{-2} \, \text{m s}^{-1}$$

Hence an upward velocity of water of 7.35 cm s^{-1} will carry away particles smaller than 300 μm diameter and allow larger particles to sink.

(b) (i) 200 μm; (ii) 600 μm; (iii) water plus particles less than 300 μm diameter; (iv) particles less than 300 μm diameter; (v) particles greater than 300 μm diameter; (vi) classifier; (vii) water.

SAQ 20 Your flowsheet may differ in detail, but the general sequence should be the same as in Figure 85.

SAQ 21 In a classifier the particles are relatively widely dispersed, and settle at rates proportional to their diameters. The turbulent counter-current prevents any aggregation of small particles. Hence slow-sinking small particles are separated from fast-sinking larger particles.

In a thickening tank there is no counter-current, and the particles are very densely packed in suspension. Small particles aggregate, and the aggregates settle at similar rates to the larger particles. Aggregation can be intensified by slight agitation and chemical flocculating agents.

SAQ 22 All seven statements are true: (a) See Figure 72; (b) see Figure 73; (c) see Figure 72; (d) see Figure 73; (e) this is in practice the same question as (c); (f) see Figure 72; (g) see Figure 81 (the mixture should yield iron at about 1 020 K).

SAQ 23 Molten slag is run off from the surface of molten metal (Section 8.2) and hence should readily rise to the surface and flow freely. Thus it should have a low melting point, a low viscosity and a low density.

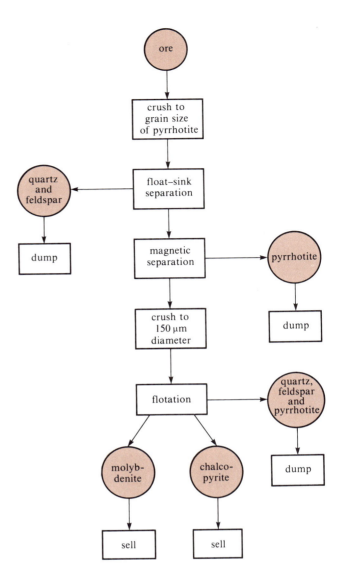

Figure 85 Answer to SAQ 20.

Acknowledgements

Grateful acknowledgement is made to the following sources for material used in Block 3 Part II:

Figure 2 from W. C. Peters, *Exploration and Mining Geology,* © 1978 John Wiley and Sons Ltd, reprinted by permission; *Figure 3* reproduced by permission of National Academy of Sciences, Washington; *Figure 5* from K. C. Dunham in *Quarterly Journal of the Geological Society of London,* no.123, 1976; *Figure 6* aerial photograph on p.13 © Her Majesty the Queen in Right of Canada, reproduced from the collection of the National Air Photo Library, with permission of Energy, Mines and Resources of Canada; *Figure 7a* from F. F. Sabins, 'Engineering applications of remote sensing' in *Geology, Seismicity and Environmental Impact,* D. E. Moran (ed.) The Association of Engineering Geologists, 1973; *Figure 7b* provided through courtesy of the Jet Propulsion Laboratory, California Institute of Technology, Pasadena, California; *Figures 10 and 12* from A. W. Rose *et al., Geochemistry in Mineral Exploration,* © Academic Press Inc. (London) Ltd, reprinted by permission; *Figure 16* from D. H. Griffiths and R. F. King, *Applied Geophysics for Engineers and Geologists,* Pergamon Press, 1965; *Figure 23* from M. B. Dobrin, *Introduction to Geophysical Prospecting,* 2nd edn. 1960, © McGraw-Hill Book Company, reprinted by permission; *Figure 29* from Heiland, Tripp and Wantland in AIME *Transactions,* vol.164, 1945, Society of Mining Engineers; *Figure 30* reprinted by permission from *Nature,* vol.272, no.5653, copyright © 1978 Macmillan Journals Ltd., *Figure 43* photo by John Simmons; *Figure 45* photo by Dr Bradford Washburn, Rapho Agence de Presse Photographique; *Figures 61 and 68* from A. Grierson and M. P. Jones, *Materials and Technology,* vol.3, Longman, 1970; *Figure 63* courtesy of BBC Hulton Picture Library.

The Course Team acknowledge with gratitude Dr A. N. McLaurin, BP Minerals International Ltd., (Block 3 Part II assessor).